PHASE

TRANSITIONS

PHASE
TRANSITIONS

Ricard V. Solé

PRINCETON UNIVERSITY PRESS
Princeton & Oxford

Copyright © 2011 by Princeton University Press

Published by Princeton University Press,
41 William Street, Princeton, New Jersey 08540

In the United Kingdom: Princeton University Press,
6 Oxford Street, Woodstock, Oxfordshire OX20 1TW

All Rights Reserved

ISBN: 978-0-691-15075-8

Library of Congress Control Number: 2011927326

British Library Cataloging-in-Publication Data is available

This book has been composed in AGaramond and Helvetica Neue

Printed on acid-free paper ∞

press.princeton.edu

Printed in the United States of America

1 3 5 7 9 10 8 6 4 2

THIS BOOK IS DEDICATED
TO THE MEMORY OF
BRIAN GOODWIN,

who helped me to find the road less taken

CONTENTS

PREFACE

Many important problems of complexity are related in one way or another with the presence of phase transition phenomena. Most complex systems are known to potentially display a number of different patterns of qualitative behavior or *phases*. Such phases correspond to different forms of internal organization and two given phases are usually separated by a sharp boundary, and crossing such a frontier implies a change in system-level behavior. Many of these transitions reveal important features of the system undergoing them. A good model of a complex system should be able to predict the presence of such phases and their implications. Some of these changes can be catastrophic, and understanding them is crucial for the future of biodiversity or even our society.

The existence of different phases that the same system can display has been well known in physics for centuries. That water freezes and becomes solid or boils to become a gas are part of our daily experience with matter. And yet, a close examination of these events immediately triggers deep questions: Why are these changes sharp? Why not simply a continuous transformation? What drives these changes at the microscopic level? How much of these details are needed to understand the phenomenon at the large scale? The presence of phase transitions in ecology, economy, epidemics, or cancer, just to cite a few examples discussed in this book, has been a matter of analysis and theoretical work over the last decades. Examples of them are scattered over the technical literature and I have tried to put together a relevant group of examples.

This book is not a general treatise of phase transitions, which would require a rather thick volume, given the multiple aspects required to cover the field. Instead, I have chosen a particular approximation, —the so-called mean field theory—and in particular those cases where such theory allows simplification into the most basic mathematical model. Mean field theory ignores some crucial aspects of reality, such as the local nature of interactions among elements or the role played by fluctuations. In removing these components we can formulate the simplest possible model.

Many people have contributed to this book. I would like to thank colleagues that have crossed my path during this long journey, particularly those at the Santa Fe Institute, where much of this book was written. I am also indebted to the members of the Complex Systems Lab for stimulating discussions. The book contains a list of references to help the reader further explore the problems presented here and that are beyond the limited scope of the mean field picture. The list is not exhaustive but contains books and papers that helped me understand the problems and look beyond them. I hope this book becomes part of that list.

PHASE

TRANSITIONS

1

PHASE CHANGES

1.1 Complexity

By tracking the history of life and society, we find much evidence of deep changes in life forms, ecosystems, and civilizations. Human history is marked by crucial events such as the discovery of the New World in 1492, which marked a large-scale transformation of earth's ecology, economics, and culture (Fernandez-Armesto 2009). Within the context of biological change, externally driven events such as asteroid impacts have also triggered ecosystem-scale changes that deeply modified the course of evolution. One might easily conclude from these examples that deep qualitative changes are always associated with unexpected, rare events. However, such intuition might be wrong. Take for example what happened around six thousand years ago in the north of Africa, where the largest desert on our planet is now located: the Sahara. At that time, this area was wet, covered in vegetation and rivers, and large mammals inhabited the region. Human settlements emerged and developed. There are multiple remains of that so-called Green Sahara, including fossil bones and river beds. The process of desertification was initially slow, when retreating rains changed the local climate. However, although such changes were gradual, at some point the ecosystem collapsed quickly. The green Sahara became a desert.

Transitions between alternate states have been described in the context of ecology (Scheffer 2009) and also in other types of systems, including social ones. Complex systems all display these types of phenomena (at least as potential scenarios). But transitions can also affect, sometimes dramatically, molecular patterns of gene activity within cells, behavioral patterns of collective exploration in ants, and the success or failure of cancer or epidemics to propagate (Solé et al. 1996; Solé and Goodwin 2001). When a given parameter is tuned and crosses a threshold, we observe change in a system's organization or dynamics. We will refer to these different patterns of organization as *phases*. The study of complexity is, to a large extent, a search for the principles pervading self-organized, emergent phenomena and defining its potential phases (Anderson 1972; Haken 1977; Nicolis and Prigogine 1977, 1989; Casti 1992a, b; Kauffman 1993; Cowan et al. 1994; Gell-Mann 1994; Coveney and Highfield 1995; Holland 1998; Solé and Goodwin 2001; Vicsek 2001; Mikhailov and Calenbuhr 2002; Morowitz 2002; Sornette 2004; Mitchell 2009). Such transition phenomena are collective by nature and result from interactions taking place among many interacting units. These can be proteins, neurons, species, or computers (to name just a few).

In physics, phase changes are often tied to changes between order and disorder as temperature is tuned (Stanley 1975; Binney et al. 1992; Chaikin and Lubensky 2001). Such *phase transitions* typically imply the existence of a change in the internal symmetry of the components and are defined among the three basic types of phases shown in figure 1.1. An example of such transition takes place between a fluid state, either liquid or gas, and a crystalline solid. The first phase deals with randomly arranged atoms, and all points inside the liquid or the gas display the same properties. In a regular (crystalline) solid, atoms are placed in the

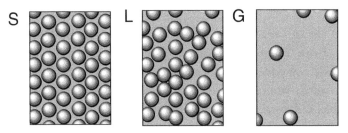

Figure 1.1. Three standard phases of matter: solid (S), liquid (L) and gas (G). Increasing a given external parameter (the control parameter) such as temperature, we can increase the degree of disorder. Such disorder makes molecules fluctuate around their equilibrium positions, move around them, or just wander freely.

nodes of a regular lattice. In the gas phase (at high temperature) kinetic energy dominates the movement of particles and the resulting state is homogeneous and isotropic. All points are equivalent, the density is uniform, and there are essentially no correlations among molecules. In the liquid phase, although still homogeneous, short-distance interactions between molecules leads to short-range correlations and a higher density. Density is actually the fundamental difference distinguishing these two phases. Finally, the ordered arrangement observed at the solid phase is clearly different in terms of pure geometry. Molecules are now distributed in a highly regular way. Crystals are much less homogeneous than a liquid and thus exhibit less symmetry.

However, beyond the standard examples of thermodynamic transitions between these three phases, there is a whole universe. Matter, in particular, can be organized in multiple fashions, and this is specially true when dealing with so-called soft matter. But the existence of different qualitative forms of macroscopic organization can be observed in very different contexts. Many are far removed from the standard examples of physics and chemistry. And yet, some commonalities arise.

1.2 Phase Diagrams

Phase changes are well known in physics: boiling or freezing are just two examples of changes of phase where the nature of the basic components is not changed. The standard approach of thermodynamics explores these changes by defining (when possible) the so-called equation of state (Fermi 1953), which is a mathematical expression describing the existing relations among a set of state variables (i.e., variables defining the state of a system). For an ideal gas, when no interactions among molecules need be taken into account,[1] the equation reads

$$pV = nRT \qquad (1.1)$$

where p, V, and T indicate pressure, volume, and temperature, respectively. Here R is the so-called gas constant (the same for all gases) and n the amount of substance (in number of moles). This equation is valid for a pure substance, and as we can see, it establishes a well-defined mathematical relation between p, V, T, and n. Given this equation, only three independent variables are at work (since the fourth is directly determined through the state equation). From this expression, we can plot, for example, pressure as a function of V and T:

$$p(V, T) = \frac{nRT}{V} \qquad (1.2)$$

which describes a continuous surface, displayed in figure 1.2a. For a given amount of perfect gas n, each point on this surface defines the only possible states that can be observed. Using this picture, we can consider several special situations where a given variable, such as temperature, is fixed while the other two can be

---

[1] In this approximation, the molecules are considered point objects. However, in spite of this extreme oversimplification, the model can be used in a number of applied contexts.

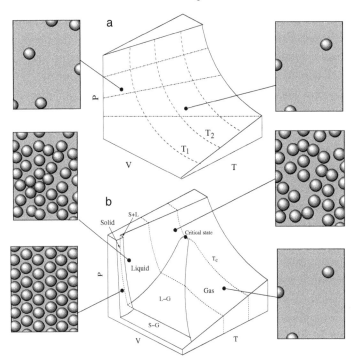

Figure 1.2. The equation of state describes the relations between the three key variables (p, V, T) that allow definition of a surface. In (a) the corresponding surface $p(V, T)$ for an ideal gas is shown. This is obtained from the equation of state $pV = nRT$, which gives a continuous surface associated to a single gas phase. In reality (b) things are much more complex and the surface is discontinuous. This second plot corresponds to a substance (such as water) that expands when it solidifies. At high temperatures it is steam. Here kinetic energy is larger than the potential energy. But once T is decreased, nonlinear changes occur, as is evident from the discontinuous shape of the surface.

changed. Fixing a given temperature T_1, we can see, for example, that pressure decays with volume following an inverse law, that is, $p(V; T_1) = nRT_1 / V$, which allows defining an isothermal process as the one moving on this line. The curve is called an *isotherm*. By using different temperatures, we can generate different isotherms (which are in fact cross sections of the previous surface). Similarly, we can fix the volume and define another set of curves, now given by $p(T) = BT$ with $B = nR/V$. The important idea here is that all possible states are defined by the equation of state and that in this case all possible changes are continuous.

The previous equations are valid when we consider a very diluted gas at high temperature. However, in the real world, transitions between different macroscopic patterns of organization can emerge out of molecular interactions[2] (figure 1.2b). Different phases are associated with different types of internal order and for example when temperature is lowered, systems become more ordered. Such increasing order results from a dominance of cohesion forces over thermal motion. Different combinations of temperature and pressure are compatible with different phases. An example is shown in figure 1.3, where the phase diagram for water is displayed[3] involving liquid, solid, and gas phases. Two given phases are separated by a curve, and crossing one of these curves implies a sharp change in the properties of the system. We say that a *first order* transition occurs and, in this case, each phase involves a given type of organization though no coexistence between phases is allowed. The melting of a solid (such as ice) or the boiling of a liquid (such as water) is a daily example, where both solid and liquid are present at the melting point.

There is a special point in the previous diagram that appears when we follow the liquid-gas boundary curve and defines its limit. This so-called critical point describes a situation where there is in fact no distinction between the two phases. Moreover, we

---

[2] This diagram is in fact just a small subset of the whole spectrum of possible phases displayed by water, which exhibits many other exotic phases (Stanley et al. 1991; Ball 1999).

[3] Let us notice that this is not a completely typical phase diagram for a liquid, since the fusion curve has positive slope.

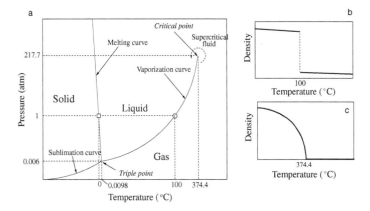

Figure 1.3. Phase diagram for water. Three basic phases are defined, separated by well-defined critical curves. A special path is indicated around the critical point that allows moving from one phase to the other (here from gas to liquid) without crossing any critical curve. Figures (b) and (c) illustrate the two different types of phase transitions that can occur.

can see that the lack of a boundary beyond the critical point makes possible *continuous* movement from one phase to the other, provided that we follow the appropriate path (such as the one indicated in figure 1.3 with a dashed curve). The presence of this point has a crucial relevance in understanding the nature and dynamics of many natural and social phenomena.

1.3 Interactions Make a Difference

Phase transitions have been shown to occur in many different contexts (Chaikin and Lubensky 1995; Stanley et al. 2000; Solé and Goodwin 2001). These include physical, chemical, biological, and even social systems. In figure 1.4 we illustrate several examples, including those from molecular, behavioral, and cellular biology and cognitive studies. These are systems spanning many orders of magnitude in their spatial embodiment, and the

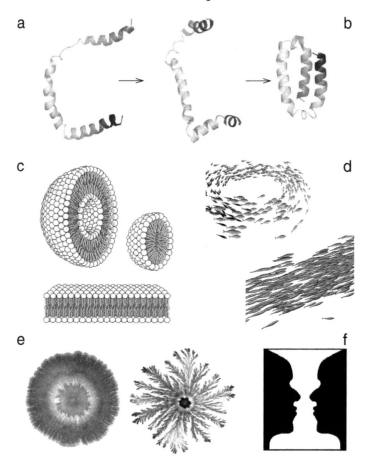

Figure 1.4. Phase transitions can be identified in many different contexts. So (a–b), proteins, for example, starting from a linear chain (a) fold into three-dimensional objects (b) able to perform given functions. Amphiphilic molecules (such as lipids) spontaneously form complex spatial structures in water. Depending on the concentration of molecules, different patterns are obtained (c) including layers and closed vesicles (image from http://en.wikipedia.org/wiki/Micelle). Swarm behavior in fish schools (d) can be modeled by means of different types of agent-based approaches. By changing the right parameters, we can see transitions between different types of collective motion
(continued)

nature of the transitions is very different in terms of its functional and evolutionary relevance, but all of them have been described by means of simple models.

Proteins exhibit two basic classes of spatial organization. Either they are folded in three-dimensional space or remain unfolded (as a more or less linear chain). The change from one state to the other (a–b) takes place under critical conditions (temperature or even the presence of other molecules). Our second example deals with a special and important group of molecules, so-called amphiphiles, which possess both hydrophilic and hydrophobic properties. Typically, there is a hydrophilic head part (see figure 1.4c) that gets in touch with water molecules, whereas the opposing side, the so-called hydrophobic tail, avoids interacting with water. As a consequence, sets of amphiphilic molecules will tend to form self-organized, spatial structures that minimize the energy of interaction (Evans and Wennerström 1999; Mouritsen 2005). By depending on the relative concentrations of water and amphiphiles, different stable configurations will be observed, defining well-defined phases (such as layers or vesicles).

Our third and fourth examples involve behavioral patterns of interactions among individuals within groups of animals, as these occur with a fish school (figure 1.4d) or, at a lower scale of organization, among cells in a cell culture (figure 1.4e). In the first case, interactions involve repulsion, attraction, the tendency to maintain the same movement as neighboring individuals or to remain isolated. This is easily modeled using computer simulations and mathematical models. By tuning the appropriate parameters, we

---

Figure 1.4 (continued). Bacterial colonies can display markedly different spatial patterns of growth (e) when the concentration of a given nutrient is tuned (modified from Cohen et al., 1996). And finally, (f) phase transitions occur in our brains as we shift from one image (a vase) to the alternate one (two faces) while looking at an ambiguous picture. (Pictures courtesy of Jacques Gautrais)

can observe different types of collective motion and colony-level patterns (Vicsek et al. 1995; Toner and Tu 1998; Czirok et al. 1999; Camazine et al. 2001; Mikhailov and Calenbuhr 2002; Theraulaz et al. 2003; Schweitzer 2003; Sumpter 2006; Garnier et al. 2007; Gautrais et al. 2008). Growing bacterial cell populations in a Petri dish with variable concentrations of a critical nutrient also demonstrate that different forms of colony organization emerge once given thresholds are crossed (Ben Jacob 2003; Ben Jacob et al. 2004; Kawasaki et al. 1997; Matushita et al. 1998; Eiha et al. 2002; Wakano et al. 2003). The final case (figure 1.4f) involves cognitive responses to ambiguous figures (Attneave 1971) and a dynamical example of transitions among alternate states (Ditzinger and Haken 1989). Our brain detects one of the figures (the two faces) or the other (the vase), and both compete for the brain's attention, which alternates between the two "phases" (Kleinschmidt et al. 1998).

In this book we present several examples of phase transitions in very different systems, from genes and ecosystems to insect colonies or societies. Many other examples can be mentioned, including chemical instabilities (Nicolis et al. 1976; Feinn and Ortoleva 1977; Turner 1977; Nitzan 1978), growing surfaces (Barabási and Stanley 1995), brain dynamics (Fuchs et al. 1992; Jirsa et al. 1994; Kelso 1995; Haken 1996, 2002, 2006; Steyn-Ross and Steyn-Ross 2010), heart rate change (Kiyono et al. 2005), immunology (Perelson 1989; de Boer 1989; Tomé and Drugowich 1996; Perelson and Weisbuch 1997; Segel 1998), galaxy formation (Schulman and Seiden 1981, 1986) cosmological evolution (Guth 1999; Linde 1994), computation (Landauer 1961; Huberman and Hogg 1987; Langton 1990; Monasson et al. 1999; Moore and Mertens 2009), language acquisition (Corominas-Murtra et al. 2010), evolution of genetic codes (Tlusty 2007), politics and opinion formation (Lewenstein et al. 1992; Kacperski and Holyst 1996; Weidlich 2000; Schweitzer 2002, 2003; Buchanan 2007), game theory (Szabo and Hauert

2003; Helbing and Yu 2009; Helbing and Lozano 2010), and economic behavior (Krugman 1996; Arthur 1994, 1997; Ball, 2008; Haldane and May 2011) to cite just a few.

1.4 The Ising Model: From Micro to Macro

In the previous sections we used the term *critical point* to describe the presence of a very narrow transition domain separating two well-defined phases, which are characterized by distinct macroscopic properties that are ultimately linked to changes in the nature of microscopic interactions among the basic units. A critical phase transition is characterized by some *order parameter* $\phi(\mu)$ that depends on some external *control parameter* μ (such as temperature) that can be continuously varied. In critical transitions, ϕ varies continuously at μ_c (where it takes a zero value) but the derivatives of ϕ are discontinuous at criticality. For the so-called first-order transitions (such as the water-ice phase change) there is a discontinuous jump in ϕ at the critical point.

Although it might seem very difficult to design a microscopic model able to provide insight into how phase transitions occur, it turns out that great insight has been achieved by using extremely simplified models of reality. In this section we introduce the most popular model of a phase transition: the *Ising model* (Brush 1967; Stanley 1975; Montroll 1981; Bruce and Wallace 1989; Goldenfeld 1992; Binney et al. 1993; Christensen and Moloney 2005). Although this is a model of a physical system, it has been used in other contexts, including those of membrane receptors (Duke and Bray 1999), financial markets (Bornhodlt and Wagner 2002; Sieczka and Holyst 2008), ecology (Katori et al. 1998; Schlicht and Iwasa 2004), and social systems (Stauffer 2008). Early proposed as a simple model of critical behavior, it was soon realized that it provides a powerful framework for understanding different phase transitions using a small amount of fundamental features.

The Ising model can be easily implemented using a computer simulation in two dimensions. We start from a square lattice involving $L \times L$ sites. Each site is occupied by a *spin* having just two possible states: -1 (down) and $+1$ (up). These states can be understood as microscopic magnets (iron atoms) pointing either north or south. The total magnetization $M(T)$ for a given temperature T is simply the sum $M(T) = (1/N) \sum_{i=1}^{N} S_i$ where $N = L^2$. Iron atoms have a natural tendency to align with their neighboring atoms in the same direction. If a "down" atom is surrounded by "up" neighbors, it will tend to adopt the same "up" state. The final state would be a lattice with only "up" or "down" units. This defines the *ordered* phase, where the magnetization either takes the value $M = 1$ or $M = -1$. The system tries to minimize the energy, the so-called Hamiltonian:

$$\mathcal{H} = -\frac{1}{2} \sum_{\langle i,j \rangle} J S_i S_j \qquad (1.3)$$

where $J > 0$ is a coupling constant (the strength of the local interactions) and $\sum_{\langle i,j \rangle}$ indicates sum over nearest neighbors.[4]

The model is based on the following observation. If we heat a piece of iron (a so-called ferromagnet) to high temperature, then no magnetic attraction is observed. This is due to the fact that thermal perturbations disrupt atomic interactions by flipping single atoms irrespective of the state of their neighbors. If the applied temperature is high enough, then the atoms will acquire random configurations, and the global magnetization will be zero. This defines the so-called disordered phase. The problem thus involves a conflict between two tendencies: the first toward order, associated to the coupling between nearest atoms, and the second toward disorder, due to external noise.

---

[4] This model can be generalized by including an external field h_i acting on each spin and favoring the alignment in a preferential direction. This can be done by adding a term $-\sum_i h_i S_i$ to the Hamiltonian.

We can see that pairs of units with the same value will contribute by *lowering* the energy whereas pairs with opposite values will *increase* it. Thus changes resulting from spin-spin interactions will occur in the direction of reducing the energy of the system by aligning spins in the same direction.

The reader can check that the minimal energy state (the *ground state*) is either all spins up or down (in both cases, we have $S_i S_j = 1$ for all pairs). However, as soon as we consider thermal fluctuations, the system will be able to explore different configurations. We will be interested in those more likely to occur. In this context, it can be shown that the probability $P(\{S_i\})$ of observing a given set $\{S_i\}$ of spins (a *state*) is:

$$P(\{S_i\}) = \frac{1}{Z} \exp\left(-\frac{E(\{S_i\})}{kT}\right) \tag{1.4}$$

where Z is the normalization $Z = \sum_{\{S_i\}} \exp(-E(\{S_i\}/kT)$ also known as the *partition function*.[5] Here $\exp(-E(\{S_i\})/kT)$ is the so-called Boltzmann factor. As we can see, increasing T has a strong effect: the exponential decay becomes slower and for high T the factor approaches 1 for all energies. The flatness implies that all states are equally likely to be observed. A typical configuration will be a mixed set of randomly distributed spins.

1.5 Monte Carlo Simulation

The model runs start with a completely disordered set, where each point in the lattice can take up or down states with the same probability. Using a fixed temperature T and coupling constant J, the new state is obtained by means of a simple set of rules to be

---

[5] The partition function captures all the relevant information required in order to recover all the thermodynamic functions, including free energy or entropy, among others.

applied iteratively using a so-called Monte Carlo method (Landau and Binder 2000):

1. Choose a random spin S_i.
2. Compute the energy change ΔE associated to the flip $S_i \rightarrow -S_i$.
3. Generate a random number $0 \le \xi \le 1$ with a uniform distribution.
4. If $\xi < \exp(-\Delta E / kT)$ then flip the spin.[6] If not, leave it in its previous state.
5. Repeat [1].

The energy difference ΔE weights the likelihood of the transition $S_i \rightarrow -S_i$ of taking place. The probability for such transition, will be

$$W(S_i \rightarrow -S_i) = \exp\left(-\frac{\Delta E}{kT}\right) \qquad (1.5)$$

if $\Delta E < 0$ and $W(S_i \rightarrow -S_i) = 1$ otherwise. As defined, we can see that the temperature is randomizing the system's behavior (the higher T, the more close to a coin toss we get). This stochastic set of rules is known as the Metropolis algorithm (Landau and Binder 2000). By iterating these rules, we decrease the energy of the system and thus approach the most likely state. In figure 1.5 we plot the result of computer simulations using a 100×100 lattice.

We should not be surprised if this caricature of reality fails to reproduce anything quantitatively relevant. But let us simulate it in two dimensions using the temperature as a tuning parameter. Each time step, the magnetization $M(T, t)$ is calculated. Starting from a random initial condition, we can compute the order parameter, defined here as the average $\langle M(T) \rangle$ for different

---

[6] The k parameter is the so-called Boltzman constant, which for convenience we take as unity in our simulations.

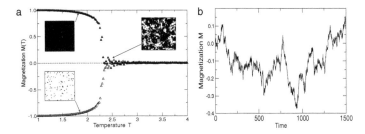

Figure 1.5. Dynamics and spatial patterns in the two-dimensional Ising model. Here (a) shows that the transition takes place at $T_c = 2.27$. Here a 100×100 system has been used. Starting from a random initial condition, for each temperature the system is run over 5×10^2 steps and afterward M is averaged over 10^3 steps. Two different sets of initial conditions have been used, starting with more (filled triangles) and less (open triangles) than half-up spins at time zero. In (b) we display a time series of M for T_c.

temperatures. Specifically, the model is run for a large number of steps and the average is computed over the last τ steps (here we use $\tau = 5 \times 10^4$, see figure 1.5). Afterward $M(T)$ is plotted against the control parameter T. In other words, we plot the average:

$$\langle M(T, t) \rangle_\tau = \frac{1}{\tau} \sum_{t=t^*}^{t^*+\tau} M(T, t) \tag{1.6}$$

Two sets of initial conditions have been used to generate the two curves shown here. Open and filled triangles indicate an initially higher and lower number of *up* spins at time zero, respectively. A sharp transition takes place at a critical temperature T_c. For low and high temperatures, we find what we expected (inset figure 1.5a): a homogeneous (ordered system) with $|M| \approx 1$ and a random ($M \approx 0$) system, respectively. But at criticality the system displays wide fluctuations in M, with an average value of zero.

This is shown by the time series of $M(T, t)$ displayed in figure 1.5b. As we can clearly appreciate, wide fluctuations are present, far from what would be expected in a system at (or

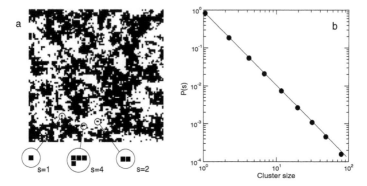

Figure 1.6. Clusters in the Ising model. A cluster involving nearest spins having the same orientation (say "up") is defined by a set of spins that are nearest neighbors. Three examples are shown in (a) for clusters of size $s = 1$, $s = 2$ and $s = 4$. Here we define the members of a cluster by considering only four nearest positions. The cluster size distribution at criticality (i.e., for $T = T_c$) is displayed in (b) in a log-log plot where data are binned using powers of two. The slope τ is close to two.

close) to equilibrium. These fluctuations are a characteristic of critical points, but we will ignore them in our treatment of phase transitions. Similarly, we could also measure the distribution of "islands" of different sizes, composed by groups of spins with the same orientation (figure 1.6a). If such distribution is computed, it is found that most clusters are composed of only one element, followed (with much smaller numbers) by two-spin clusters, and so forth. The shape of this distribution is a power law. Specifically, if $P(s)$ indicates the frequancy of clusters of size s, this probability decays as:

$$P(s) \sim s^{-\tau} \tag{1.7}$$

In a log-log plot, this will approach a straight line (figure 1.6b). Power laws are associated to many different complex systems in nature and society (Solé and Goodwin 2001; Ball 2008) and indicate the presence of large fluctuations.

1.6 Scaling and Universality

We have used the Ising model as an illustration of how the conflict between order and disorder leads to long-range correlations in equilibrium systems, where a well-defined stationary distribution of energy states exists. Other models of equilibrium phenomena display similar behavior. In spite of the presence of a characteristic scale of local interactions, structures at all scales emerge. But the interest of the model resides in its *universal* character and illustrates the astonishing power of model simplicity (Chaikin and Lubensky 1995; Stanley et al. 2000). One can determine a number of different quantities from the model, such as fractal dimensions or correlation lengths. Near the critical point, the system displays scale invariance: If we observe such system at a spatial scale r, it looks the same when a larger or smaller scale is used.[7] The behavior of these quantities near critical points can be fully characterized by a set of numbers, so-called critical exponents (Stanley 1972; Goldenfeld 1992). For example, if we measure the magnetization M close to criticality, it is found that

$$M \sim (T - T_c)^{-\beta} \tag{1.8}$$

where $\beta(d = 2) = 1/8$ and $\beta(d = 3) = 0.325$ in two and three dimensions, respectively, to be compared with the experimental result for ferromagnetic materials, $\beta_e \approx 0.316 - 0.327$. Similarly, the correlation length ξ scales as

$$\xi \sim |T - T_c|^{-\nu} \tag{1.9}$$

---

[7] Formally, the same statistical patterns are observed when a transformation $r \to \lambda r$ is performed (for $\lambda > 0$).

where now $\nu(d=2)=1$ and $\nu(d=3)=0.630$, to be compared to the real value $\nu_e \approx 0.625$. The correlation length measures the characteristic distance over which the behavior of one element is correlated with (influenced by) the behavior of another. Its divergence at T_c implies that two distant points influence each other and is associated to the emergence of very large clusters, as shown in figure 1.6.

These exponents have been shown to be basically independent on the special traits of both interactions and components.[8] The exponents predicted by the model are exact: they fit experimental results almost perfectly. Actually, other very different physical systems (such as fluids) displaying critical phase transitions have been shown to behave exactly as ferromagnets (Yeomans 1992). The irrelevance of the details becomes obvious when we check that the exact lattice geometry or taking into account interactions beyond nearest neighbors does not modify the system behavior. Although the value of T_c is system dependent, the values of the critical exponents appear to be rather insensitive to microscopic details, thus indicating that the properties displayed by critical points are rather universal. This allows the association of each type of transition with a different *universality class*. The number of such classes is limited and thus in principle we can classify different phenomena as members of different classes of universality.

1.7 Mean Field Ising Model

The set of rules defining the Ising model are simple, so much so that it might seem an easy task to predict how the system will behave. This is far from true. As discussed at the beginning of this chapter, complexity arising from simple interacting units

---

[8] The concept of universality was originally developed by Leo Kadanoff and establishes that near critical points the details of what happens at the small scale are largely irrelevant.

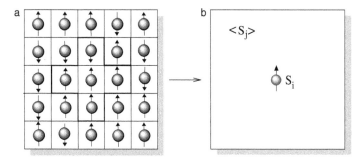

Figure 1.7. Mean field approach to the Ising model. In (a) a small lattice of spins is shown, showing both up and down states. Although each spin interacts only with a reduced number of neighbors (eight in this case) the mean field approximation (b) reduces to a much simpler scenario. A given, arbitrary spin S_i would be affected by an average field described in terms of the global average spins in the system.

is not simple nor reducible. Solving the two-dimensional Ising model was an extraordinary task, finally achieved by Lars Onsager in 1944. Among other things he predicted an exact critical temperature $T_c = 2/\log(1 + \sqrt{2}) \approx 2.27$ (Onsager 1944). The three-dimensional model was never solved exactly. It has been suggested that no such solution actually exists (Istrail 2000).

In this book we consider the simplest approximation that can be made toward understanding phase transitions: the so-called mean field theory, which is based on a number of strong assumptions and ignores the local nature of interactions. We know that each element in the Ising model (or in a real solid) interacts with just a few neighbors. We will ignore this and assume that each unit "feels" all the others as some kind of global field (figure 1.7). This field is nothing but the average value of all other spins. In this way, we ignore the spatial and temporal fluctuations that *we know* are present in the real counterpart. This is certainly a very strong assumption and as a result it often fails to reproduce some essential quantitative features. Nevertheless it captures the *logic* of the system and allows one to predict whether phase changes are likely to occur.

Let us illustrate the approach by solving the Ising model (figure 1.6). Consider an arbitrary spin in the system, S_i. The average value of this spin can be obtained from $\langle S_i \rangle = \sum_{S_i = +1, -1} S_i P(S_i)$ where $P(S_i)$ indicates the probability that the spin has a $+1$ or -1 state. Using the Boltzmann distribution, we have

$$P(S_i) = \frac{\exp\left[-\beta\left(J S_i \sum_j S_j\right)\right]}{\sum_{S_i = -1, +1} \exp\left[-\beta\left(J S_i \sum_j S_j\right)\right]} \tag{1.10}$$

and the average spin value now reads:

$$\langle S_i \rangle = \frac{\exp\left[-\beta\left(J \sum_j S_j\right)\right] - \exp\left[\beta\left(J \sum_j S_j\right)\right]}{\exp\left[-\beta\left(J \sum_j S_j\right)\right] + \exp\left[-\beta\left(J \sum_j S_j\right)\right]} \tag{1.11}$$

the previous expression can be transformed using the equivalence[9] $\tanh(\mu x) = [\exp(-\mu x) - \exp(\mu x)]/[\exp(-\mu x) + \exp(\mu x)]$ which for our case gives:

$$\langle S_i \rangle = -\tanh\left[\beta\left(J \sum_j S_j\right)\right] \tag{1.12}$$

This function is known to be bounded by two assymptotic limits, namely $\tanh(\mu x) \to \pm 1$ for $x \to \pm\infty$.

Let us look at the following sum: $H_i \equiv \left(\sum_j J S_j\right) S_i$ which is nothing but the (local) energy of interaction associated to the S_i spin. Now let us make the mean field assumption: replace the

---

[9] The hyperbolic tangent *tanh*(x) and other so-called hyperbolic functions are the analogs of the standard trigonometric functions, with several desirable properties, and widely used in many areas of physics and mathematics.

previous spins by their average values, namely $\langle S_j \rangle \equiv \sum_j S_j / N$, meaning that the exact set of states surrounding S_i is replaced by the average over the lattice. Now we have

$$H_i = \left(\langle S_j \rangle \sum_j J \right) S_i \qquad (1.13)$$

If the number of neighbors is q (four in two dimensions, eight in three, etc.) we can write $\sum_j J = J q$ and the average value of the spin is now:

$$\langle S_i \rangle = \tanh \left[\beta \left(q J \langle S_j \rangle \right) \right] \qquad (1.14)$$

Since we defined the magnetization M as the average over spins, and using $\beta = 1/kT$ we have a mean field equation for the magnetization in the Ising model:

$$M = \tanh \left[\left(\frac{qJ}{kT} \right) M \right] \qquad (1.15)$$

where we used $M \equiv \langle s_i \rangle = \langle s_j \rangle$. For convenience (as we will see below) we will use the notation $T_c \equiv qJ/k$. In figure 1.8a we plot the function $y_1 = \tanh(M)$ for different values of $\mu \equiv T_c / T$ which acts as an inverse of the temperature. We also plot the linear function $y_2 = M$. The potential equilibrium states of our system would correspond to the intersections between these two curves, that is, $y_1 = y_2$. As μ is increased (and thus the temperature reduced) the steepness of the nonlinear function y_1 increases close to $M = 0$. At high T (low μ) only one intersection is found, corresponding to the zero magnetization state. Once we have $T < T_c$ three intersections occur.

If we are interested in studying the behavior of this system close to the phase transition, where M is (on average) small, we can use the expansion $\tanh x \approx x - x^3/3 + \cdots$ By ignoring all terms of order larger than three, we obtain an approximate expression for

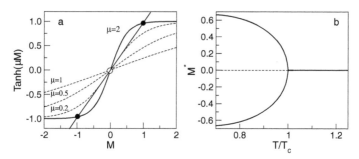

Figure 1.8. The mean field theory for the Ising model correctly predicts a phase transition close to a given critical temperature T_c. In (a) we show the shape exhibited by the function $y = \tanh(\mu M)$ against the magnetization M for different μ values. The line $y = M$ is also shown, together with the intersections with the curves, indicating solutions to the equation $M = \tanh(\mu M)$. In (b) the resulting transition curves are shown.

the magnetization:

$$M = \left(\frac{T_c}{T}\right) M - \frac{1}{3}\left(\frac{T_c}{T}M\right)^3 \tag{1.16}$$

Which values of $M(T)$ are compatible with this equation? One solution is the trivial, zero magnetization state, that is, $M(T) = 0$, and the other possible values follow $1 = (T_c/T)-(T_c/T)^3/3M^2$. Two symmetric solutions are obtained, namely:

$$M_\pm(T) = \pm\sqrt{3}\left(\frac{T}{T_c}\right)\left[1 - \frac{T}{T_c}\right]^{1/2} \tag{1.17}$$

which exist only if $T < T_c$. These solutions are shown in figure 1.7b using the reduced temperature T/T_c as our control parameter. As we can see, the mean field model properly captures the key behavior of the magnetic phase transition. Several estimations of relevant thermodynamic quantities can be derived from this approximation (see Christensen and Moloney 2005).

Since the mean field approach ignores spatial and temporal correlations (and fluctuations) it must somewhat fail to be a correct description of real systems. As a consequence, predicted scaling laws and other quantitative measurements will depart from the observed.[10] However, this theoretical approach gives in general the correct qualitative predictions of the phase diagrams, particularly as we increase the range of the interactions. We can imagine this as follows: in a d-dimensional square lattice, the number of nearest neighbors is $2d$, which means that, as d grows (i.e., as more neighbors are included) each element exchanges information with a larger fraction of the system. As a consequence, the effects of spatial degrees of freedom and fluctuations become less and less important. In fact, statistical physics analyses have shown that there is a *critical dimension* d_c above which fluctuations become unimportant and the MF approach becomes a correct description (Goldenfeld 1992).

In summary, as the number of potential interactions grows, mean field predictions become more and more exact. Moreover, as pointed out by Yeomans (1992) the mean field approximation provides a feasible approach to complicated problems and, in many cases, is the only one. Moreover, it often provides the first step toward more accurate models. For all these reasons, this is the path taken here.

1.8 Nonequilibrium Transitions

Is the Ising model our model of reference for complex systems? I think it provides an elegant example of how global complexity

---

[10] The previous formalism can be generalized by using a continuous approximation where the local states are replaced by a field $\phi(x)$ now x indicating the coordinates in a continuum space. The so-called field theory approximation allows us to explore the effects of spatial degrees of freedom and is the cornerstone of the modern theory of critical phenomena (Kardar 2007).

emerges out from local interactions. It also illustrates the nontrivial phenomena arising close to critical points. But the hard truth is that most systems in nature and society are *far from equilibrium* and no energy functions are at work (Hinrichsen 2006). In particular, equilibrium systems obey the so-called detailed balance condition. Specifically, if the system is described in terms of a set of possible states, $\mathcal{A} = \{\alpha\}$ such that $\alpha \in \mathcal{A}$ occurs with a probability P_α, for each pair $\alpha, \alpha' \in \mathcal{A}$, a transition probability $\omega(\alpha \to \alpha')$ can be defined. The detailed balance condition is defined as follows:

$$P_\alpha \omega(\alpha \to \alpha') = P_{\alpha'} \omega(\alpha' \to \alpha) \qquad (1.18)$$

which can be easily interpreted: the probability flows from one state to another, in both directions, and cancels the other out.

Nonequilibrium systems are not determined (solely) by external constraints, and violate the detailed balance condition. As a given control parameter is changed, a given stationary state can become unstable and be replaced by another. Since detailed balance is not operating, their macroscopic properties are no longer obtainable from a stationary probability distribution. Besides, in systems out of equilibrium, no free energy can be defined. In equilibrium systems, the phase diagram can be determined on the basis of the free energy but this approach becomes useless when we move far from equilibrium.

The difference is particularly dramatic when dealing with systems involving *absorbing states* (Hinrichsen 2000; Marro and Dickman 1999). Absorbing configurations can be reached by the dynamics but cannot be left. As an example, a population can go extinct once a given threshold is reached, but there is no way to get out from extinction. Not surprisingly, extinction and collapse will be two important examples of nonequilibrium transitions explored in this book.

2

STABILITY AND INSTABILITY

2.1 Dynamical Systems

Our first basic approach to the analysis of phase changes in complex systems involves the consideration of a fundamental property: stability. Our perception of the stable and the unstable is filtered by our perception of change. But in general terms it can be said that stable things are those that do not vary over long time scales, whereas instability is tied to change. Complex systems maintain a stable organization thanks to a balance between internal order (which needs to be preserved) and flexibility (adaptability).

A mechanical analog of the basic ideas to be discussed below is displayed in figure 2.1. Here we have a surface on top of which one can place marbles. Although a marble could be left on top of a peak and remain there, any small perturbation would force the ball to move away (and thus the peaks are unstable). Instead, the bottom of a valley is a stable point, since small perturbations of the marble only change its position for a while, until it goes back to the bottom again (to the stable equilibrium state). Marginal points are represented by locations on flat domains, where a small movement to a neighbor location just leaves the ball in the same place, without moving away or back to the initial position.

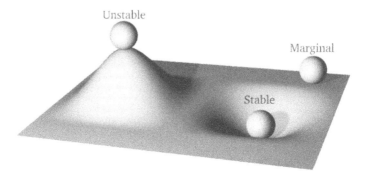

Figure 2.1. The three types of (linear) stability, through a simple mechanical analogy. Here three marbles are at equilibrium at three different points in the landscape (From Solé and Bascompte 2006).

The simplest dynamical system that we can consider is described by a *linear* differential equation. Imagine a population of bacteria replicating at a given rate. Let us indicate as μ the rate of replication of individual bacteria. Clearly, the more bacteria we have in our Petri dish, the faster the production of new bacteria. If we indicate by means of $x(t)$ the number of bacteria at a given time t, after a small time Δt we would have:

$$x(t + \Delta t) = x(t) + \mu x(t)\Delta t \qquad (2.1)$$

This discrete equation just tells us that the process is incremental and proportional to the existing population. The last term on the right-hand side indicates that the longer the interval Δt, the larger the number of newborn bacteria.

Although this equation is useful, it is rather limited: we would like to know how many bacteria are expected to be found after a given time, assuming that the initial number $x(0)$ at a given zero time is known. In order to find this, we need to transform the previous difference equation.

The time interval Δt is an arbitrary one, and we can make it as small as we want. By rearranging the previous linear equation, we have:

$$\frac{x(t + \Delta t) - x(t)}{\Delta t} = \mu x(t) \qquad (2.2)$$

Using $x = x(t)$, and considering the limit when $\Delta t \to 0$, it can be transformed into a *differential equation*:

$$\frac{dx}{dt} = \mu x \qquad (2.3)$$

where dx/dt is the derivative of x against t, measuring the speed at which the population grows with time. As we can see, this speed is proportional to the current population, increasing linearly with x. For this reason, we define this model as a *one-dimensional linear dynamical system*.

2.2 Attractors

The linear differential equation can be used to find the explicit form of the $x(t)$ time dependent population dynamics. Solving the differential equation in this case is an easy task. We just separate the variables, that is:

$$\frac{dx}{x} = \mu dt \qquad (2.4)$$

and perform the integration in both sides. Assuming that at time $t = 0$ the initial population was x_0 and that the population size at any arbitrary time $t > 0$ is indicated as $x(t)$ we have:

$$\int_{x_0}^{x(t)} \frac{dx}{x} = \int_0^t \mu dt \qquad (2.5)$$

which gives the solution

$$x(t) = x_0 e^{\mu t} \qquad (2.6)$$

The predicted behavior indicates that an explosive (exponential) growth will take place, with rapidly increasing numbers of individuals as time proceeds.

A very different scenario is provided by the linear decay defined by the differential equation:

$$\frac{dx}{dt} = -\mu x \qquad (2.7)$$

where the speed of change is now a negative linear function of population size. This equation is also common in different contexts, such as radioactive decay:[1] if x indicates the number of atoms of a given radioactive material, they disintegrate at a given rate μ. The corresponding solution is now

$$x(t) = x_0 e^{-\mu t} \qquad (2.8)$$

where x_0 would represent the total number of unstable atoms at the beginning of the experiment. In this case, when we consider the limit at infinite time, we obtain

$$\lim_{t \to \infty} x(t) = \lim_{t \to \infty} x_0 e^{-\mu t} = 0 \qquad (2.9)$$

since asymptotically all atoms will eventually decay.

Sometimes, we can reduce the complexity of a given problem involving several variables to a single-equation model. Let us consider an example that has some interesting features. The problem is illustrated in figure 2.2, where we display a schematic diagram of two connected compartments where a gas is confined. Initially, all molecules are placed in the left compartment. The two boxes are connected through a valve of fixed diameter that is open at time zero. Once open, the gas particles can flow between the two compartments. What is the dynamics of gas exchange? We know that eventually the molecules will be equally distributed between both compartments and such a result can be easily shown by means of an appropriate use of mass conservation.

---

[1] Radioactive decay is a stochastic process: any given radioactive atom of a given isotope can disintegrate at any time with a given probability. However, taken together and considering a large number of atoms, the statistical behavior of the material is well described by a deterministic linear model.

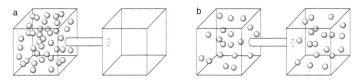

Figure 2.2. An example of a simple physical system where conservation laws permit reduction of its mathematical description. Here a two-compartment system is considered, with a gas initially located in the left compartment (a) that diffuses through time until it reaches equilibrium (b). The system is closed and the total number of particles is thus conserved.

Let us call C the initial concentration of molecules in the first compartment. Once the gas starts flowing after opening the valve, the concentrations in each box (indicated as x and y) will change in time, according to a *system* of two differential equations:

$$\frac{dx}{dt} = \mu(y - x) \tag{2.10}$$

$$\frac{dy}{dt} = \mu(x - y) \tag{2.11}$$

which are clearly linear.[2] The parameter μ is a so-called diffusion constant and weights how easily particles move from one compartment to the other.

Since we assume a closed system, we have $x + y = C$ and thus we have:

$$\frac{dx}{dt} = \mu(C - 2x) \tag{2.12}$$

which gives an evolution equation:

$$x(t) = \frac{C}{2}[1 - e^{-2\mu t}] \tag{2.13}$$

---

[2] In general, a system of n linear differential equations will read as $dx_i/dt = \sum_j \alpha_{ij} x_j$ for $i, j = 1, \ldots, n$ and being α_{ij} a matrix of real coefficients.

As expected from our intuition, the final state is determined by the limit:

$$\lim_{t \to \infty} \frac{C}{2}[1 - e^{-2\mu t}] = \frac{C}{2} \tag{2.14}$$

and thus the equilibrium state is $x^* = y^* = C/2$: the gas uniformly distributes over space. Any initial condition will eventually end up in this equilibrium state. The example thus illustrates one possible scenario where dimensional reduction can be obtained by means of conservation.

We must mention here that linear differential equations are seldom a good description of complex systems, but as will be shown in the next section, the previous examples will be useful in developing a general approximation to nonlinear dynamics. Moreover, although we have solved analytically the previous models, finding an explicit form for $x(t)$ is in most cases impossible. However, since we are mainly interested in the observable stationary states, we will use a different strategy of analysis. This strategy makes use of a general technique that determines the type of stability displayed by each fixed point.

2.3 Nonlinear Dynamics

In general, we will consider dynamical systems described by means of nonlinear differential equations. The general expression in our one-dimensional representation would be

$$\frac{dx}{dt} = f_\mu(x) \tag{2.15}$$

where $f_\mu(x)$ is some type of nonlinear function of x (such as x^2, x^3, $\cos(\mu x)$ or $1/(\mu+x)$). An alternative notation commonly used is:

$$\dot{x} = f_\mu(x) \tag{2.16}$$

where $\dot{x}$ indicates derivative of x over time. The function is assumed to be continuous and with continuous derivative. The μ symbol is used to remind us that in general the dynamics will depend on one or several parameters.

As an example of a well-known nonlinear system, consider the logistic growth model (Pielou 1977; Case 1999). If x indicates the population size of a given species growing on a limited set of resources, its time evolution can be described by means of the following deterministic equation:

$$\frac{dx}{dt} = f_\mu(x) = \mu x \left(1 - \frac{x}{K}\right) \qquad (2.17)$$

where μ and K are the growth rate and carrying capacity of the population, respectively. We can easily see that when x is very small compared to K it can be approximated by $dx/dt \approx \mu x$, which means that an exponential growth should be expected. However, the growth will be limited by available resources, and for $x \approx K$ (close to the limit) the last term in the right-hand side of the logistic equation will go to zero, and thus the speed of population growth will decrease to zero too. These patterns can be studied by solving the differential equation analytically.[3] If $x = x_0$ at $t = 0$, it can be shown that

$$x(t) = \frac{K}{1 + \left[\frac{K-x_0}{x_0}\right] e^{-\mu t}} \qquad (2.18)$$

and thus $x(t) \to x^* = K$ as $t \to \infty$. In other words, any initial condition $x_0 > 0$ eventually converges to the carrying capacity $x^* = K$, consistent with our intuition. In figure 2.3a we show several examples of the trajectories followed by the logistic dynamics, as described in equation (17) using $\mu = 1$ and $K = 2$ and

---

[3] Once again, we can separate variables and solve the integrals $\int dx/(x(1 - x/K)) = \int \mu dt$. This is done by decomposing the x-dependent fraction as a sum $1/(x(1 - x/K)) = A/x + B/(1 - x/K)$ and then determining the values of A and B consistently and integrating separately the two terms.

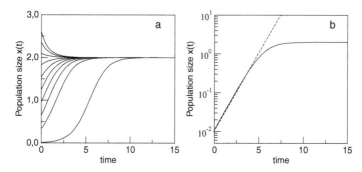

Figure 2.3. Solutions of the logistic differential equation, for $\mu = 1.0$ and $K = 2$. In (a) several initial conditions are used, all of them converging to the steady state $x^* = K$, which acts as an attractor. In (b) we show a single solution, with $x(0) = 0.01$ in linear-log scale. The dashed line indicates the exponential law $x(t) \approx x(0)e^{\mu t}$ which is valid for short times.

different initial conditions. We can see how all of them approach the carrying capacity as time proceeds, thus indicating that $x = K$ is a stable state. In figure 2.3b we show one of these trajectories (starting from $x(0) = 0.01$) in linear-log scale. In such a plot, exponential solutions appear as straight lines. As we can see, this is the case for the first part of the dynamics, as expected: The exponential solution obtained from the linear approximation $dx/dt \approx \mu x$ is also observable (dashed line) showing that the linear approximation is good for small time scales. Since all trajectories starting from $x_0 > 0$ become eventually "attracted" by x^*, we call it an *attractor* of the dynamics. This is also the case for other models, in which stable points describe the long-term behavior of the solutions.

Instead of finding the analytic solution of the differential equation (which is not possible most of the time) we can study the behavior of the solutions close to the fixed points. More precisely, we want to determine how to classify different types of fixed points in terms of their stability. In order to do so, we will study the behavior of the system close to the equilibrium points, using a linear approximation to the nonlinear dynamics. This defines the so-called linear stability analysis (LSA).

2.4 Linear Stability Analysis

The first step in LSA is to identify the set π_μ of equilibrium (fixed) points. Since equilibrium will occur whenever $dx/dt = 0$ (no time changes take place) this is equivalent to finding those x^* such that the function $f_\mu(x^*)$ is zero. The resulting set of points will be:

$$\pi_\mu = \{x^* \mid f_\mu(x^*) = 0\} \qquad (2.19)$$

From our definition, if $x(0) = x^*$, it is true that $x(\infty) = x^*$. But obvious differences appear when comparing the behavior of different fixed points. The basic idea is illustrated using again the mechanical analog with marbles shown in figure 2.1. The three of them are at equilibrium. However, a small displacement from their equilibrium state will affect them in very different ways. The *stable* equilibrium point will be recovered after some transient time (as it occurs with a pendulum). The *unstable* one will never recover its original position: the initial perturbation will be amplified and the marble will roll away. A third, apparently less interesting possibility is the so-called marginal state: on a totally flat surface, a marble displaced to another close position will just stay there.

The nature of the different points $x^* \in \pi_\mu$ is easily established by considering how a small perturbation will evolve in time (Strogatz 1998; Kaplan and Glass 1999). Consider $x = x^* + y$, where y is a small change (such as adding a small amount of individuals to the equilibrium population). Using the general form of the one-dimensional dynamics given by eqn (15) and since x^* is a constant, the dynamics of the perturbation y is obtained from:

$$\frac{dy}{dt} = f_\mu(x^* + y) \tag{2.20}$$

Assuming that y is very small, we can Taylor expand[4] the previous function:

$$\frac{dy}{dt} = f_\mu(x^*) + \left[\frac{df_\mu}{dx}\right]_{x^*} y + \frac{1}{2!}\left[\frac{d^2 f_\mu}{dx^2}\right]_{x^*} y^2 + \cdots \tag{2.21}$$

If y is very small, we can safely ignore the higher-order terms (with y^2, y^3 ...). Since $f_\mu(x^*) = 0$ (by definition) and provided that its first derivative is non-zero, we can use the linear approximation:

$$\frac{dy}{dt} = \lambda_\mu y \tag{2.22}$$

where λ_μ is defined as:

$$\lambda_\mu = \left[\frac{df_\mu}{dx}\right]_{x^*} \tag{2.23}$$

or alternatively as

$$\lambda_\mu = \left(\frac{\partial \dot{x}}{\partial x}\right)_{x^*} \tag{2.24}$$

The linear system is exactly solvable, with an exponential solution:

$$y(t) = y(0)e^{\lambda_\mu t} \tag{2.25}$$

This solution immediately provides a criterion for stability: if $\lambda_\mu < 0$, then $y \to 0$ and thus the system returns to its *stable* equilibrium. If $\lambda_\mu > 0$, perturbations grow and thus the point

---

[4] In general, a Taylor expansion of a given function about a point x^* is given by an infinite series, namely $f(x) = f(x^*) + f'(x^*)(x - x^*)/1! + f''(x^*)(x - x^*)^2/2! + \cdots$. In our notation, we have $y = x - x^*$ which gives the size of the perturbation relative to x^*.

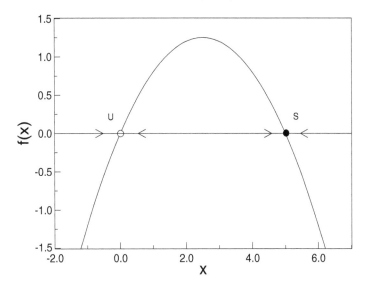

Figure 2.4. The fixed points of a given dynamical system can be easily obtained by plotting the function $f_\mu(x)$ against x and finding its intersections with the horizontal axis. Here the logistic curve is shown for $\mu = 1$ and $K = 5$. Stable and unstable fixed points are indicated as filled and open circles, respectively.

is *unstable*. Both possibilities are separated by a *marginal* state defined from $\lambda_\mu^c = 0$. Thus just looking at the sign of λ_μ we can immediately identify the presence of stable and unstable states. Using the condition $\lambda_\mu = 0$ we will also characterize the boundaries separating stable and unstable patterns of behavior.

This finding can be geometrically described by plotting the function $f_\mu(x)$ against x (figure 2.4). The intersections of f with the horizontal axis correspond to the fixed points, and the slope of the curve near them is related to their stability. At unstable points, the derivative of $f_\mu(x)$ is positive, which means positive slopes. Instead, stable points will have negative slopes. By plotting this picture, we can immediately identify how many fixed points are present and their stability.

We can illustrate these general results by means of the logistic model just described. Here we have a set of fixed points defined by $\pi_{\mu,K} = \{0, K\}$ and the stability of each equilibrium point is determined by the sign of

$$\lambda_\mu = \mu \left(1 - \frac{2x^*}{K} \right) \qquad (2.26)$$

In this case we have we have $\lambda(0) = \mu$ which is always positive (since the growth rate is a positive parameter) and thus $x^* = 0$ is unstable. For the second point we have $\lambda(K) = -2\mu$ which is always negative and K is consequently always stable, consistent with our intuition.

Stability theory, as defined here, will be our first tool in the study of mean field models of phase changes. The second ingredient has to do with stability changes: how and when a given stable state becomes unstable, being replaced by a new, qualitatively different type of state.

3

BIFURCATIONS AND CATASTROPHES

3.1 Qualitative Changes

As discussed in chapter 1, our interest here is understanding the presence of different types of behavior defining different phases. As it occurs with physical systems, phases can be defined and their boundaries characterized. The stability analysis presented in the previous chapter allows exploring changes in the behavior of complex systems in terms of changes in their stability.

We have already mentioned a few examples of phase changes in the first chapter. Here our goal is to provide a simple but useful mathematical approach to characterize the problem. Following the formalism introduced in the previous chapter, we will consider models that allow describing a given phase change in terms of one-dimensional differential equations. Such a simplification, as already discussed, implies a number of limitations. But it also helps understanding how we can capture some key aspects of complexity by properly choosing the right observable.

The key concept presented here is the idea of *bifurcation* (Jackson 1990; Nicolis 1995; Kuztnesov 1995; Strogatz 1998), a term attributed to Henri Poincaré. A bifurcation is a qualitative change in the dynamics of a given system that takes place under continuous variation (tuning) of a given parameter. Bifurcations will be the analog to phase transitions under the continuous, dynamical systems view taken here. Mathematically a bifurcation implies the emergence (and disappearance) of new solutions. In the real world, it relates to a qualitative change when, for example, information, viruses, or fires propagate (what thresholds need to be overcome) or a shift occurs in the structure of an ecosystem. An example of the latter would be the transition from vegetated to desert habitats. Moreover, some specially relevant bifurcations (see next section) imply the appearance of alternate states. Such bifurcations introduce an interesting element: the possibility of path-dependent phenomena. In these cases, relevant to both biological and technological evolution (Arthur 1994; Gould 1992), the initial condition plays a key role. Small differences in the initial state can actually lead to very different outcomes, thus introducing a role for history.

3.2 Symmetry Breaking

In many situations, a given system is forced to make a choice between two potential candidate states. These two states can be equally likely chosen, and the final choice is often decided by means of some historical contingency. Once again, the situation is illustrated by means of a mechanical analogy, shown in figure 3.1. In this case, a single marble is located on top of a ridge separating two identical valleys. Any small change in the marble's position will force it to roll down toward the nearest stable state (the bottom of the nearest valley). The direction of such initial perturbation plays a relevant role here, since only one of the two possible choices can be made. The initial symmetry is broken.

Such symmetry breaking (SB) mechanism allows us to create structures and pervades the generation of patterns in many situations in biology, chemistry, and physics (Haken 1975, 1996; Nicolis and Prigogine 1989; Strocchi 2005). The two alternative orientations of spins displayed by the Ising model (figure 1.5) is a

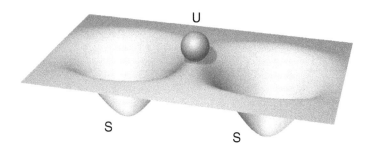

Figure 3.1. Symmetry breaking: a marble can roll down on a landscape where two alternative (stable) valleys (S) can be reached. The valleys are identical and located symmetrically from the original position of the marble, which is in an unstable state (U). After Solé and Bascompte, 2006.

good illustration of this phenomenon. By depending on the initial balance between up and down spins, the magnet evolves toward either a positive or a negative global magnetization, respectively.

The SB phenomenon is very well illustrated by the following dynamical model:

$$\frac{dx}{dt} = \mu x - x^3 \qquad (3.1)$$

Which appears in many different contexts. The set of fixed points of this system is easily determined. Together with the trivial fixed point $x^* = 0$ we have two symmetric points, namely

$$x^*_\pm = \pm\sqrt{\mu} \qquad (3.2)$$

The stability of these fixed points is determined by:

$$\lambda_\mu = \mu - 3(x^*)^2 \qquad (3.3)$$

For $x^* = 0$, we obtain $\lambda_\mu(0) = \mu$: This point is thus stable if $\mu < 0$ and unstable otherwise. For the *two* other fixed points, we have now $\lambda_\mu(x^*_\pm) = -2\mu$ which means that they both exist and are stable for $\mu > 0$. These fixed points and their stability are easily determined by drawing $f_\mu(x)$ and looking at its slope around the intersections with the x-axis (figure 3.2).

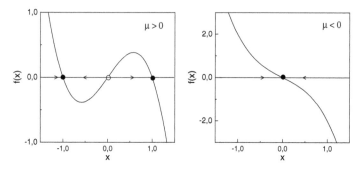

Figure 3.2. The presence and type of equilibrium points in the dynamical system $dx/dt = \mu x - x^3$. For $\mu > 0$ (left) three crossings occur between $\mu x - x^3 = 0$ and the x axis, thus indicating the presence of three fixed points, one of them unstable (open circle) and two stable ones (filled circle). For $\mu < 0$ (right) only one crossing is present, indicating a single (stable) fixed point.

In figure 3.3a we show the trajectories $x(t)$ for this system, using a negative value of the bifurcation parameter μ and starting from different initial conditions. The corresponding solutions are shown in figure 3.3b for a positive value of μ. These curves have been calculated by *numerically* integrating the differential equation using the so-called Euler's method.[1] As predicted by the LSA, solutions converge to $x^* = 0$ in the first case while they split into two different attractors for the second case.

We can integrate all the previous results by using a simple diagram. Such a *bifurcation diagram* provides a picture of the available equilibrium states and their stability as the bifurcation parameter μ is tuned. In figure 3.3c the bifurcation diagram for

---

[1] Euler's method is the simplest algorithm that allows solving of a differential equation. It is based on the discrete approximation to the differential equation already discussed in the introduction section. Given an initial condition $x(0)$ and fixing the time interval Δt (typically a small value) we compute $x(\Delta t)$ by using $x(\Delta t) = x(0) + f_\mu(0)\Delta t$, and the next value from $x(2\Delta t) = x(\Delta t) + f_\mu(\Delta t)\Delta t$ and so on.

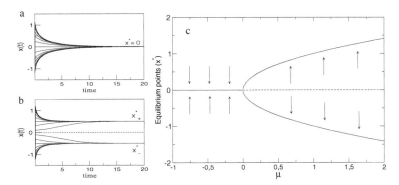

Figure 3.3. Symmetry breaking in a one-dimensional dynamical system, as defined in equation (1). In (a) we show several numerical solutions for this model using $\mu = -0.5$ starting from different initial conditions. The same is shown in (b) for $\mu = 0.5$. The corresponding bifurcation scenario in displayed in (c).

this model is shown. It is constructed by representing all the fixed points $x^* \in \pi_\mu$ for each value of μ. Stable fixed points are joined by continuous lines, whereas unstable ones are located along discontinuous lines. We indicate by means of arrows the direction of trajectories on the x-space toward stable states. The bifurcation diagram also allows to define the boundaries and behavior of the two qualitatively differentiated *phases* displayed by this system. It is worth noting the similarities between this example and the mean field treatment of the Ising model described in chapter 1.

The qualitative properties of the SB phenomenon can also be captured by a mathematical counterpart of the surface-and-marble metaphor. This is achieved by using a so-called potential function $\Phi_\mu(x)$, defined from the relation

$$\frac{dx}{dt} = -\frac{d\Phi_\mu(x)}{dx} \tag{3.4}$$

where $\Phi_\mu(x)$ would be computed from

$$\Phi_\mu(x) = -\int f_\mu(x)dx = -\int \dot{x}dx \tag{3.5}$$

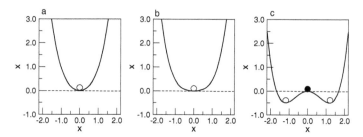

Figure 3.4. The potential $\Phi_\mu(x)$ for the system described by equation (1) using: (a) $\mu = -1.5$, (b) $\mu = -0.4$ and (c) $\mu = 1.5$. The location of the stable and unstable fixed points is indicated by means of empty and filled circles, respectively. For $\mu > 0$ two alternative states are simultaneously available, symmetrically located on both sides of $x^* = 0$. Any small perturbation of the zero state will decide the chosen equilibrium among the two, thus breaking the symmetry of the system.

It is usually said that a dynamical system that can be written as (3.4) *derives* from a potential $\Phi_\mu(x)$. The existence of a potential function will depend on the integrability of $f_\mu(x)$ but for smooth enough functions[2] is typically guaranteed. The logic of this description is clear if we notice that the same points that give $dx/dt = 0$ will be those such that $d\Phi_\mu(x)/dx = 0$ and thus corresponds to the extrema of $\Phi_\mu(x)$. Moreover, since we have $d^2\Phi_\mu(x)/dx^2 = -df_\mu(x)/dx = -\lambda_\mu$, the sign of the second derivative will be correlated with the type of equilibrium point. Specifically, the minima and maxima of $\Phi_\mu(x)$ will correspond to stable and unstable points, respectively.

For the previous example, we have

$$\Phi_\mu(x) = \frac{x^4}{4} - \mu\frac{x^2}{2} \qquad (3.6)$$

[2] This essentially means that the function has no singularities, which occurs when f and its derivatives are continuous.

and its shape is displayed in figure 3.4 for two values of μ. When the bifurcation parameter is negative, this function has a single minimum at the bottom of the valley whereas when positive the situation dramatically changes: two valleys are now present. Two alternate states are available, and the previous stable point is now at the top of a ridge. From there, tiny differences will decide the final fate of the system.

Although this book does not consider models including explicitly random perturbations (the Ising model is our exception) it is worth mentioning such effects in the context of symmetry breaking. Instead of making an analytic derivation, let us use a simple mechanical analog to illustrate the point. This is summarized in figure 3.5, where we show a marble rolling at the bottom of a surface. This would roughly correspond to the potential function of our previous model (although an additional dimension has been added) below but close to the bifurcation point. Close to this point, as shown in figure 3.4b, the bottom of the valley flattens and the flatness increases close to criticality. Now imagine our marble rolling at the bottom of this lanscape, sometimes quickened by small, external perturbations. Since the

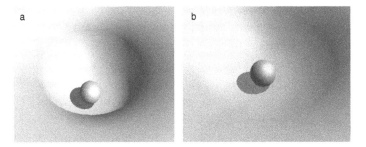

Figure 3.5. Fluctuations close to the symmetry breaking transition. Here we represent a close-up of two potential functions $\phi_\mu(x)$ with a shallow (a) and a shallower (b) bottom. The flatter the surface, the wider will be the wandering of the marble around the equilibrium point.

valley is rather flat, any small perturbation can drive the marble far away from the equilibrium point. As a consequence, the marble might spend long periods without even approaching the equilibrium state. The closer we are to the critical point, the more dramatic become these fluctuations. This has important consequences for the relaxation time required to achieve the steady state, as discussed in section 3.4.

3.3 Catastrophes and Breakpoints

The previous example illustrating the SB phenomenon is known in physics as a *second order phase transition*. In the context of the physical theory of phase transitions, such a phenomenon takes place as the control parameter μ changes and some *order parameter* (our x) moves continuously from one phase to the other. However, in some cases a slow change in the control parameter leads to a sudden shift in the system's state or, as it is sometimes dubbed, a *catastrophe* (Thom 1972; Zeeman 1977; Poston and Stewart 1978; Deakin 1980; Gilmore 1981; Arnold 1984; Jackson 1991). This type of phenomenon occurs (as discussed in chapter 1) in the water-ice transition at sea level pressure: as temperature (our control parameter) decreases below the freezing temperature, water displays a transition to the solid state. At the microscopic scale, the liquid water with all its molecules dancing around experiences a deep geometrical change, being that now the molecules are arranged in an ordered manner.

Rapid shifts from a given state to a new one, separated by a finite gap, are known to occur in many different contexts, from ecology (Sheffer et al. 2001; Scheffer and Carpenter 2003) to climate (Foley et al. 2003) and social dynamics (Ball 2006). The changes can be dramatic and an example is illustrated in figure 3.6a, where we return to the example discussed in the introduction to this book. What is displayed here is a time series of the amount of dust measured in ocean core sediments over

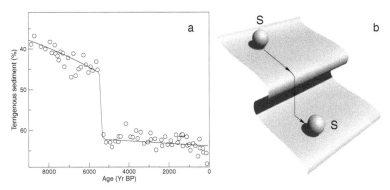

Figure 3.6. Catastrophic shifts: In (a) we display the time evolution of the dust measured from sediments at a given location (redrawn from Scheffer et al. 2001). This measure provides a surrogate of the area covered by terrigenous (wind-eroded) soils. About fifty-five hundred years ago, a sudden drop took place (see text). The mechanical analog of such catastrophic shifts is given in (b).

a long period of time, spanning nine thousand years to the present. We can see that the amount of dust decreased slowly until a well-defined drop occurred around fifty-five hundred years ago. The measurement was obtained near the African coast in the northern hemisphere, and the shift is actually associated with a large-scale, abrupt change (perhaps occurring over the course of a century) from a vegetated to a desert state in the geographical area that is today known as the Sahara and Sahel deserts. Despite that the shift from vegetation to desert was sharp, it was associated with continuous variations in the amount of solar radiation (deMenocal et al. 2000). In other words, smooth changes in external drivers of vegetation dynamics led to a sudden change in the ecosystem organization. This situation might be repeating itself currently in semiarid ecosystems (Kefi et al. 2007; Scanlon et al. 2007; Solé 2007; Kefi 2008; see also chapters 12 and 13). A mechanical analog of this phenomenon is shown in figure 3.6b. Here a marble is shown on top of a folded surface. If it rolls down, it will move smoothly until the fold is reached.

Afterward, it will fall into a new area, separated from the previous by a well-defined gap. Is there a mathematical counterpart to this mechanical analogy? An example will help us see that indeed this is the case.

Consider now a slight generalization of our previous model, given by:

$$\frac{dx}{dt} = \alpha + \mu x - x^3 \qquad (3.7)$$

Due to the presence of the new parameter α, the analysis of the system becomes more complicated: the set of fixed points π_μ is now obtained from $\alpha + \mu x^* - (x^*)^3 = 0$ and the general form of them is rather involved. We know, however, that either one or three solutions will be possible. Instead of finding analytically these points, we will use a shortcut. The fixed points are found at the intersection of the curve

$$y_1 = \mu x - x^3 \qquad (3.8)$$

and the constant line $y_2 = -\alpha$. As it happened with our previous system, we have two very different qualitative behaviors. For $\mu < 0$ a single fixed point is obtained, whereas for positive values (an example is shown in figure 3.7a) there is a domain of α values, $-\alpha_c < \alpha < +\alpha_c$ where three simultaneous fixed points are present. As we move the constant line y_2, we can see that it crosses the y_1 curve in different ways. For some values only one crossing is allowed (giving a single stable point) whereas for a domain of values three crossings occur. The impact of these folded lines on the system's dynamics is illustrated in figure 3.7b, where we show that the change in the equilibrium states as α is linearly and continuously tuned (inset c). Starting from $\alpha = 0.5$ and the corresponding value of x_+^*, we can see that once $\alpha < \alpha_c = 0.25$ a dramatic shift takes place, with the state jumping to the lower branch x_-^*. This perfectly illustrates the basic principle that for these types of systems a slow change of parameters does not always produce a slow dynamical response.

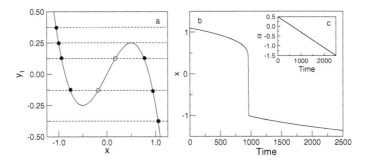

Figure 3.7. Catastrophes in a simple nonlinear model. In (a) we plot the two curves $y_1 = \mu x - x^3$ (for $\mu = 0.75$) and $y_2 = -\alpha$ (dashed lines) whose intersections define the possible fixed points. In (b) we can see how a continuous change in a control parameter can trigger a sudden shift (b). Here α is varied linearly over time (inset c).

In order to determine the boundaries of the three-state domain, which now depend on two parameters, we need to compute α_c. First, let us find the value $x = x_c$ where the extrema of $y_c = \mu x_c - x_c^3$ occurs. It is obtained from the condition $dy/dx = 0$ and is given by:

$$x_c = \pm \sqrt{\frac{\mu}{3}} \tag{3.9}$$

and the corresponding y_c will be the value α_c, namely

$$\alpha_c = \pm \frac{2\mu}{3} \sqrt{\frac{\mu}{3}} \tag{3.10}$$

Using the previous results, we can determine the domain $\mathcal{D}$ in the (μ, α) parameter space where three fixed points exist. This is now shown in figure 3.8a, where the two possible phases are indicated. We also show (figure 3.8b) the corresponding surface with the location of the fixed points $x^* = x^*(\mu, \alpha)$, as defined by the implicit function $\alpha + \mu x^* - (x^*)^3 = 0$. We can appreciate a remarkable shape in the folded surface, not very different from the one we already displayed in our mechanical analog. What is the importance of this shape?

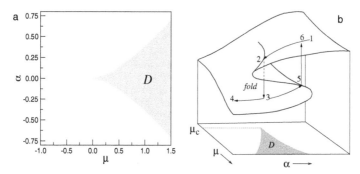

Figure 3.8. Domains of multistability in dynamical systems. In (a) we show the associated domain of the (μ, α)-parameter space for the previous example. The cusp shown in gray indicates the domain where three solutions are possible. In (b) the corresponding folded surface associated with the possible fixed points is also shown (see text for details).

A very important consequence of this folded surface is the presence of a memory effect or *hysteresis*. The surface $x^*(\alpha, \mu)$ is shown in figure 3.8b, where we start from a given stationary state 1 on the surface and continuously tune α in such a way that the system changes continuously on the surface as the control parameter is varied. After reaching point 2 the state jumps into the lower part of the surface (at 3) and if we keep reducing α a new state is reached under continuous modifications (4). However, if we start now on the lower branch again from 3 and now increase α, we do not jump back to 2 but instead move on to the lower branch until the fold is reached again, this time at point 5. A new jump occurs, but the trajectory followed is totally different.

The sudden changes associated to these transitions can also be seen by looking at the corresponding potential function, which now reads:

$$\Phi_\mu(x) = -\alpha x - \mu \frac{x^2}{2} + \frac{x^4}{4} \qquad (3.11)$$

By plotting the potential using different values of the bifurcation parameters, we can easily appreciate an important difference with respect to the potential functions involved in SB phenomena.

3.4 Critical Slowing Down

An important property displayed by systems exhibiting symmetry-breaking bifurcations is the presence of a divergence in the relaxation time close to criticality. Specifically, as the critical point is approached, the characteristic time needed to reach the equilibrium state rapidly grows. As we have already mentioned, when the system exhibits stochastic fluctuations, the variance of these fluctuations close to criticality will increase and actually exhibit a marked maximum when looking to experimental systems (Haken 1977, 1996; Nicolis and Prigogine 1989). Here we derive the time required to approach the stationary state asuming deterministic dynamics.

We can illustrate this phenomenon by using the previous model (3.1) and calculating the transient time T_μ required to go from a given initial condition $x(0)$ to a final state $x(T_\mu)$ as a function of the control parameter μ. In general, such transient time will be obtained from:

$$\int_0^{T_\mu} dt = \int_{x(0)}^{x(T_\mu)} \left[f_\mu(x) \right]^{-1} dx \qquad (3.12)$$

which needs to be split into two terms, separating the two domains at each side of $\mu_c = 0$. We will have then:

$$T_\mu^-(\epsilon) = \int_{x(0)}^{\epsilon} \frac{dx}{x(\mu - x^2)} \qquad T_\mu^+(\epsilon) = \int_\epsilon^{\sqrt{\mu}-\epsilon} \frac{dx}{x(\mu - x^2)} \qquad (3.13)$$

for the $\mu < \mu_c$ and the $\mu > \mu_c$ domains, respectively. The integrals are performed using different types of initial and final x-states. For the first part, where the fixed point is $x^* = 0$ we start from a given small value $x(0)$ and end up close to x^*,

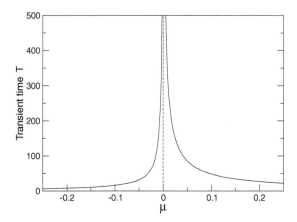

Figure 3.9. Divergence of transient times in the symmetry breaking problem described by the dynamical system $\dot{x} = \mu x - x^3$. Here we plot the analytic result obtaind in the text for $T_\mu^-(\epsilon)$ (left curve) and $T_\mu^+(\epsilon)$ (right), using $\epsilon = 0.001$ and $x(0) = 0.05$.

specifically $x = \epsilon \ll 1$. Similarly, the transient for the second half is computed by starting from $x(0) = \epsilon$ to $\sqrt{\mu} - \epsilon$.

The resulting calculations give our analytic estimates of the transient times:

$$T_\mu^-(\epsilon) = \frac{1}{2\mu} \ln \left[\frac{\epsilon^2(\mu - x(0)^2)}{x(0)^2(\mu - \epsilon^2)} \right] \qquad (3.14)$$

and

$$T_\mu^+(\epsilon) = \frac{1}{2\mu} \ln \left[\frac{(\mu - \epsilon^2)(\sqrt{\mu} - \epsilon^2)}{\epsilon^2(\mu - (\sqrt{\mu} - \epsilon)^2)} \right] \qquad (3.15)$$

These two curves are displayed in figure 3.9 against the control parameter μ. From both sides of the critical point, the transient rapidly increases, becoming infinite at $\mu_c = 0$. For μ values very close to μ_c we have in fact $T_\mu(\epsilon) \sim 1/\mu$ which gives a hyperbolic decay around the critical point.

The key implication of this behavior is that a dynamical signature of criticality is a sharply marked increase in the relaxation time exhibited by systems close to the phase transition. Such increase has been reported from different experimental studies, from brain (Haken 1996) to ecosystem dynamics (see chapters 12–13). Actually, it can be used as a warning signal of a close critical transition and thus as a potential indicator of future changes.

3.5 Multiple Broken Symmetries

The bifurcations considered in this book are all related to changes leading to different types of behavior. In the real world, they correspond to transitions between alternative states and emergence of new possible attractors, and can occur at very different scales (from a cell to a society). But in fact we need to keep in mind that in some cases the system undergoes a cascade of bifurcations. An example of this situation is biological development (Goodwin 1994). Through the process of embryo development, millions of events happen, from cell divisions to cell death and differentiation into different cell states. All these events are associated with nonlinear dynamical processes where gene-gene and cell-cell interactions play a role. This was early envisioned by the great biologist Conrad Hal Waddington (Waddington 1957; Slack 2003) who thought about the problem of development as a multistep series of bifurcations (figure 3.10a). In Waddington's picture, it is possible to visualize the whole process as a ball (the embryo's state) rolling down through a multidimensional space. Every fork in this landscape would correspond to a bifurcation.

What controls the shape of the landscape? Clearly, different actors will be involved, including genes and their interactions. Such elements would define a complex network both shaping and constraining the dynamics of development on the landscape, as schematically indicated in figure 3.10b. This picture is a powerful one and has been useful in understanding the evolution of novelty (Barton et al. 2007). Moreover, it is also an interesting metaphor

a

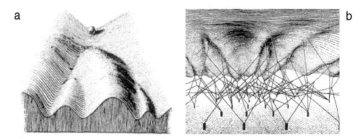

b

Figure 3.10. Multiple paths in a complex, nonlinear dynamical system, as illustrated in (a) by the so called *epigenetic landscape* of Waddington (see text). Here the underlying dynamical system is the process of development of an organism. The state of this process is indicated by the position of the ball. The process of development would correspond to the ball rolling down through this surface. The shape of the landscape would be controlled (b) by the genes involved in the process, defining a network of interacting units.

for other types of complex systems: broken symmetries pervade multiple aspects of complexity, including the generation of spatial structures, memory, and behavior (Anderson 1972; Palmer 1982; Hopfield 1994; Livio 2005).

4

PERCOLATION

4.1 Systems Are Connected

What makes us label a collection of objects a "system"? The most obvious answer is that there is some kind of link between different objects such that it makes sense to consider them as part of a larger structure. If a given set of species, genes, or individuals contains elements that appear somehow isolated from the majority, we will probably discard them as part of the global structure. A system is thus, roughly speaking, a set of items such that we can find a path connecting any two of them.

Once we agree on the previous minimal requirement, new questions emerge. How do the elements forming a given system get connected? How does connectivity affect their behavior? Answering these questions in a general way provides a broad view of the problem. In order to do this, let us consider a simple model that illustrates a powerful concept (Sahimi 1994; Stanley et al. 1999; Stauffer and Aharony 1999; Hinrichsen 2000; Christensen and Moloney 2005). Imagine you have an $L \times L$ square grid where each patch is either empty or occupied by a piece of flammable material. This is usually called a "tree" and thus the set of trees scattered throughout our lattice can also be dubbed a "forest." Imagine that trees are introduced in any given place with

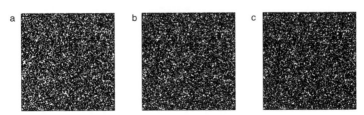

Figure 4.1. Three lattices of 150×150 sites are shown, with different densities of occupied sites (here indicated as black squares): (a) $p = 0.55$, (b) $p = 0.60$, (c) $p = 0.65$. Each site can be occupied with the same probability p, and the resulting spatial pattern is almost the same except for the effective number of occupied sites.

some probability p. At low p levels, trees will appear isolated, whereas high occupation probabilities will crowd the system, with each tree surrounded by several others. In figure 4.1 we show three examples of different levels of tree densities.

Although the density of trees increases linearly with p (we will have pL^2 trees on average) the global connectivity of the system is far from linear. Let us define the neighborhood of a given patch as the set of four nearest patches. This is the so-called von Neumann neighborhood: Trees are connected only to their four neighbors and if a tree gets burned, the fire can propagate only to those four locations (if trees are present). It seems clear that if a fire starts at some given location, it will hardly propagate at low p values and instead it will burn most trees at high p's. Our intuition suggests that the probability of a fire spreading over a large area would increase somehow linearly with p. Is this the case?

4.2 Percolation Thresholds

In order to answer the last question, let us formulate it in more quantitative terms. Suppose we burn all trees at the bottom line of our lattice and ask what the probability is that the fire will reach the upper limit of the lattice. When such an event takes

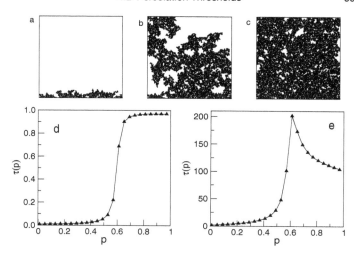

Figure 4.2. Burning trees in the previous figure: here at time zero all trees at the bottom line of each plot in the previous figure burn. The front propagates and burns further trees. For $p = 0.55$ (a) the fire goes extinct rapidly, whereas for $p = 0.65$ (c), it spreads to almost every tree in the lattice. Close to the critical point (b), a very large fire occurs, spanning the whole lattice but leaving many complex holes at all scales. The order parameter and the transient times associated with the percolation transition are shown in (d) and (e), respectively.

place, we say that the fire *percolates* the entire system. Percolation implies that there is some long path connecting points separated by a distance that is of the order of the system's size. We can simulate this process using a computer and calculate the number of burned trees for different occupation probabilities.

The results of these computer experiments are shown in figure 4.2. Despite our intuition, we can see that the fraction of trees burned is highly nonlinear. The burning actually displays a two-phase behavior. Below some threshold (figure 4.2a) the fires essentially die out very rapidly. At this subcritical phase the low density of trees makes fire propagation too difficult to succeed. Instead, once we cross the so-called percolation threshold (figure 4.2b, located at $p_c \approx 0.59$), the fire is very likely to percolate the entire lattice, and the size of the fires rapidly grows (figure 4.2c). This is rather unexpected and has a number of important consequences; the first is that we need to think of percolation as a critical phenomenon: it is in fact an example of a second-order phase transition. The second is that, in order to become a system, the minimal requirements to achieve global connectivity involve the presence of a well-defined threshold. Once such a threshold is reached, information, fires, or epidemics can easily propagate at the large scale. Determining the conditions under which percolation occurs is often equivalent to finding the boundary between two well-defined phases. Such phases describe on one hand decoupled subsystems unable to propagate signals and perturbations and, on the other, coherent structures where information is exchanged and processed globally.

This transition can also be described in terms of a control and an order parameter, as with the Ising model in chapter 1 (see figure 1.5a). Here the control parameter is p, whereas the order parameter $\Omega(p)$ can be, for example, the fraction of burned sites. Specifically, we run many simulations starting from random initial conditions (different trees will burn at the bottom of our lattices) and calculate in the end the size of the fire. In figure 4.2d we display the result of our numerical simulations for a $L = 10^2$ lattice using 10^3 runs. The result is familiar to us: a second order (critical) transition occurs near $p_c = 0.59$, where fires start to successfully propagate. Just increasing p a little beyond this point will affect most elements. The sharpness of this transition increases dramatically as the system size is increased. For very large L, we would see a marked change between an almost-zero-order parameter for all $p < p_c$ to a non-zero value after the critical point is reached.

Two important comments need to be made. First, the complexity of the pattern observed at criticality is clearly larger than the ones obtained for both $p > p_c$ (very homogeneous, random) and $p < p_c$ (homogeneous, flat). At criticality, multiple scales are involved, as illustrated by the presence of islands of patches of all sizes, displaying a fractal structure. The complexity of this structure can be measured in many ways, but a simple, indirect measure is provided by the transient time $\tau(p)$ required in order to complete the burning process. In figure 4.2e the largest transient time observed for our simulations is shown. For subcritical values of p, the fire dies out very rapidly, as expected, whereas for high p values, it is close to L, also consistent with our expectations. However, a marked peak appears around the critical point: it takes a long time to burn the complex structure shown in figure 4.2b. This is due to the complex, nested geometry of the percolation cluster.

The second point is related to the different nature of this transition (and most of those considered in this book) in relation to the Ising model. The percolation process described above is an example of a so-called absorbing phase transition (Hinrichsen 2000). These types of transitions are characterized by the presence of a stationary state that can be reached by the system but from which the system cannot escape. This is the most important class of *nonequilibrium phase transitions*, and as we mentioned in chapter 1, they violate the detailed balance condition. Such a class of behavioral pattern does not allow a connection with equilibrium thermodynamics or define energy functions, as in the Ising model.

4.3 Percolation Is Widespread

The percolation transition appears in many different contexts (Stauffer and Aharony 1999; Christensen and Moloney 2005). An interesting example is provided by the so-called sol-gel transition (Pollack 2001). Let us imagine a given set of complex molecules each having f different active centers (also called functional groups). These can be seen as different parts of the

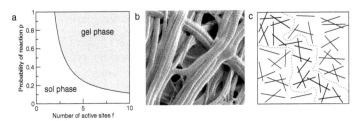

Figure 4.3. Sol-gel phase transition. In (a) the mean-field parameter space is shown, with the two phases (sol/gel) separated by the critical line $p_c = 1/(f - 1)$. In (b) an example of collagen fibers under the He-Ion electron microscope is shown (image courtesy of Zeiss Nanotechnology). Figure (c) shows a schematic representation of the percolation network formed in collagen solutions (Redrawn from Forgacs et al. 1991).

molecule that can react with other molecules, forming a bond and thus a larger macromolecular structure. Once a given pair of molecules has reacted, each one has $f - 1$ free centers which can also react. If p is the probability that a given bond gets formed, what is the expected number of bonds for a given randomly chosen molecule? It is easy to see that the average number is simply $p(f - 1)$. But now this provides a critical condition for percolation to occur: if $p(f - 1) < 1$ then on average it is likely that no further links will be made, whereas if $p(f - 1) > 1$, then at least one link will be expected. The critical boundary is thus given by:

$$p_c = \frac{1}{f - 1} \tag{4.1}$$

The resulting separation of the (p, f) parameter space into two phases is shown in figure 4.3a.

This situation appears in a number of different contexts involving polymers (Atkins 2000; de Gennes 1976). It has also been identified in some key molecular processes associated with morphogenesis (Forgacs et al. 1991; Forgacs and Newman 2007).

In some types of tissues (such as mesenchyme)[1] the morphological changes and structure displayed are largely dependent on the properties of the network of fibers that are scattered within the extracellular matrix where cells are embedded. A particularly important component is collagen.[2] Collagen undergoes an assembly process generating macromolecular fibrils and, under the appropriate conditions, macroscopic fibers. Forgacs and co-workers showed that once a critical density of fibers is reached (figure 4.3b), percolation takes place (Forgacs et al. 1991; Forgacs 1995).

4.4 Bethe Lattices

The previous derivation of the percolation point for the sol-gel example provides an estimate to the critical conditions under which the system becomes connected. What else can be said? A significant quantity is the expected size of the largest connected set of elements, the so-called giant component (see also chapter 5). Specifically, although $p_c(f)$ gives a necessary condition for the existence of a connected system, it is clear that not all elements will be connected through some path. But we can calculate the fraction of elements belonging to the giant component under a mean field approach. This approximation is based on the so-called Bethe lattice (Stauffer and Aharony 1992) and provides a simple (mean field) model of percolation. This type of lattice has a tree structure (figure 4.4a) with no loops that has been used in epidemics as a simple illustration of disease propagation.

Let us consider the nodes of this lattice as occupied or free and assume that disease or fire can propagate through from site

---

[1] This is a loosely organized connective tissue displaying a viscous consistency. It contains a network of fibers and a class of cells known as fibroblasts.

[2] This is the single most abundant molecule in the human body, making up as much as 6 percent of our body weight.

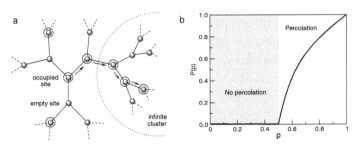

Figure 4.4. Percolation on Bethe lattices. In (a) we show an example of the tree structure of a Bethe lattice. Starting from the center moving outward, each node has a fixed number of outgoing links. In (b) the phase transition, as predicted from mean field theory arguments, is shown.

to site if and only if both sites are both occupied. The branching is assumed to be directional, which is to say, although each node has z links (the so-called coordination number), only $z - 1$ of them can actually propagate the signal. If p is the probability of occupation of a site, the average number of sites that can propagate the epidemics will be $p(z - 1)$. The critical condition for propagation defining the percolation threshold will be the same as before, namely $p_c(z - 1) = 1$.

Assuming that the lattice is infinite, we can also compute the percolation probability $P(p)$ that, given one infected site, the disease will propagate (percolate) to infinity. For $p < p_c$ we have $P(p) = 0$, but for $p > p_c$ the specific form of $P(p)$ must be derived. If we label as C_∞ an infinite (percolating cluster) and call Q the probability that a randomly chosen site $s_i \notin C_\infty$, it is not difficult to see that:

$$Q = P[s_i \notin C_\infty] = (1 - p) + p Q^{z-1} \qquad (4.2)$$

The first term on the right-hand side corresponds to the probability that the node is empty. The second term is the probability that s_i is occupied and none of the neighbors belongs to C_∞.

For $z = 3$, the previous equation gives two solutions: $Q = 1$ and $Q = (1 - p)/p$. From them the percolation probability will be:

$$P(p) = p(1 - Q^z) = p\left[1 - \left(\frac{1-p}{p}\right)^3\right] \qquad (4.3)$$

and it is shown in figure 4.4b. A rapid increase in the percolation probability is observed at $p_c = 0.5$. Actually, we can find the approximated form of $P(p)$ close to p_c by using the Taylor expansion:

$$P(p) \approx P(p_c) + \left(\frac{dP(p)}{dp}\right)_{p_c}(p - p_c)$$
$$+ \frac{1}{2!}\left(\frac{d^2P(p)}{dp^2}\right)_{p_c}(p - p_c)^2 + \cdots \qquad (4.4)$$

which gives, ignoring higher-order terms, $P(p) \approx 6(p - p_c)$. It can be shown that, the larger the value of z, the steeper the transition curve. This is consistent with the presence of larger numbers of alternative paths able to sustain propagation.

Other quantities can be derived using apprpriate mean field arguments (Christensen and Moloney 2005). As an example, consider the problem of finding the average of cluster sizes (defined in terms of sets of connected, occupied nodes). It can be shown that it is a function $\chi(p)$ given by

$$\chi(p) = \frac{p_c(1 + p)}{p_c - p} \qquad (4.5)$$

for $p < p_c$. As we can see, we have again a divergence close to p_c, that is, $\chi(p) \sim (p_c - p)^{-1}$. This result is consistent with our previous finding: close to percolation thresholds, the likelihood of finding a connected path (i.e., chains of connected elements within clusters) rapidly grows.

Percolation and bifurcation analysis can be understood as two different forms of detecting and characterizing nonequilibrium phase transitions under mean field approximations. In the next chapters, different case studies will be considered and specific analytic tools will be used, although in many cases both percolation and bifurcation theory could be used.

5

RANDOM GRAPHS

5.1 The Erdos-Renyi Model

As discussed in chapter 4 (on percolation), an interesting phenomenon that seems to pervade many facets of complexity involves the critical requirements that separate a connected from a disconnected system. In our examples we used spatially extended systems (lattices) showing that, for a threshold level of local connectivity, a signal could propagate through the system spanning its entire size. A simpler version of this idea does not consider space and is thus free from spatial correlations: it provides the simplest scenario for percolation within a set of connected elements under mean field assumptions.

Our example is based on a graph $\Omega = (V, E)$, defined by means of a set of N vertices (or nodes)

$$V = \{v_1, v_2, \ldots, v_N\} \qquad (5.1)$$

and a set of L edges (or links)

$$E = \{e_1, e_2, \ldots, e_L\} \qquad (5.2)$$

connecting pairs of vertices. For each node $v_i \in V$ we can define its degree k_i as the number of edges connecting v_i with other nodes in the graph. The average degree $\langle k \rangle$, also indicated as z, is simply

$$\langle k \rangle = \frac{1}{N} \sum_{i=1}^{N} k_i \qquad (5.3)$$

The simplest model of a random network, known as the Erdos-Renyi (ER) graph (Bollobas 1985), allows us to define a null model of a disordered network. Although this system is seldom found in reality, it provides an excellent case study when looking at how networks depart from homogeneity. This type of random model has been actually used in different contexts, including ecological (May 1976; Cohen 1978), genetic or metabolic (Kauffman 1962, 1993), and neural (Hertz 1991, Sompolinsky 1988, Amari 1988) networks.

5.2 Statistical Patterns

In ER graphs, every edge is present with probability p, and absent with probability $1 - p$.[1] The algorithm for building such web is simple: choose every pair of nodes, generate a random number $\xi \in (0, 1)$, and add a link if $\xi < p$.

The average number of edges $\langle L \rangle$ in these graphs is thus:

$$\langle L \rangle = \frac{N(N - 1)}{2} p \qquad (5.4)$$

where the $1/2$ factor is due to the fact that each link is shared by two edges, and accordingly, the average degree can be written as $z = 2\langle L \rangle / N \approx Np$ being valid only for large N. It can be easily shown that the probability $P(k)$ that a vertex has a degree k follows a Binomial shape,

$$P(k) = \binom{N - 1}{k} p^k (1 - p)^{N-k-1} \qquad (5.5)$$

---

[1] In fact, there is a whole *ensemble* of graphs for a single value of p, and random graph theory deals with the average properties of these ensembles.

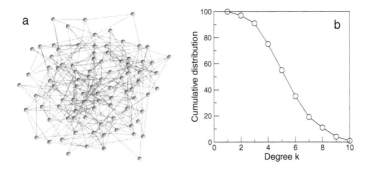

Figure 5.1. Random homogeneous networks. In (a) a typical ER graph is shown for $z = 3$. In (b) the corresponding degree distribution is shown.

and for large N and fixed z, a good description is provided by a Poisson distribution:

$$P(k) = e^{-z}\frac{z^k}{k!} \tag{5.6}$$

Moreover, the Poisson distribution has an interesting property: the variance $\sigma^2(k)$, defined as $\sigma_k^2 = \langle k^2 \rangle - \langle k \rangle^2$, can be shown to have the same value as the mean, $\langle k \rangle = \sigma_k^2$. These distributions are characterized by the average degree z: a randomly chosen vertex is likely to have a degree z or close to it. An example of the resulting network (for $N = 100$ and $z = 3$) is shown in figure 5.1, together with the corresponding degree distribution.

5.3 Percolation at Critical Connectivity

One of the most interesting properties exhibited by these graphs concerns the presence of a percolation threshold, separating a disconnected network (where no information can be exchanged among components) to a system where most (if not all) elements are linked to one another through at least one path. It is clear that increasing p in the ER model effectively increases the

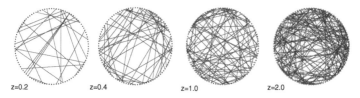

z=0.2   z=0.4   z=1.0   z=2.0

Figure 5.2. Random graphs: four examples of ER graphs are shown, for different values of the average degree $z = Np$ with $N = 100$ (here nodes are placed at equal distance along a circle). As p grows, the number of isolated elements gets reduced. Such reduction is sharp once $z_c = 1$ is crossed.

probability of two given nodes of connection through some path. Intuitively, for very small p, the graph will be fragmented into many components, most pairs being isolated or belonging to a pair of connected units. However, if p is large, most pairs of nodes will be linked and thus very short paths will be observed. Now the question is: as we increase the average number of links z, what is the probability of observing a connected (or almost connected) graph? Specifically, if S is the fraction of nodes belonging to the largest connected component (the so-called giant component), we want to know how the function $S(p)$ behaves as p is changed.[2]

Against our intuitions, and consistent with our example of a forest fire from chapter 4, such probability does not increase in a monotonous fashion with z (alternatively, with p, once N is fixed, since $z = pN$). This model displays a phase transition at a given critical average degree $z_c = 1$ (figure 5.2). At this critical point, it is said that a *giant component* forms (Bollobas 1985; an excellent presentation is given in Newman 2003a): for $z > z_c$ a large fraction of nodes are connected in a single web, whereas for $z < z_c$ the system is fragmented into (many) small subwebs.

---

[2] In all these derivations, we assume the existence of an ensemble $G_{N,p}^{ER}$ of graphs of size N and parameter p.

To see how this happens, let us follow a simple approximation that shares some points with our previous derivation using the Bethe lattice. Consider the graph Ω as the joining of a giant component S_∞ and the rest of the elements not belonging to it. Let us indicate as Q the probability that a randomly chosen element v_i does not belong to the giant component, that is, $Q = P[v_i \notin S_\infty]$ (figure 5.3). The probability of a given node of degree k not belonging to the giant component must be equal to the probability that none of its k neighbors belong to S_∞, that is, Q^k. The average value of Q must be estimated by averaging all of k, that is,

$$Q = \sum_{k=0}^{\infty} P(k) Q^k \tag{5.7}$$

For a Poissonian graph, using the Taylor expansion $\sum_k x^k / k! = e^x$, we obtain:

$$Q = e^{-z} \sum_{k=0}^{\infty} \frac{(zQ)^k}{k!} = e^{-z(Q-1)} \tag{5.8}$$

Since the size S of the giant component is $S = 1 - Q$, we have the following transcendental equation for the growth of the giant component with z:

$$S = 1 - e^{-zS} \tag{5.9}$$

We can see that for $z \to \infty$ we have $S \to 1$. Besides, we can also check that for $z = 1$ the previous equation is valid for $S = 0$. The numerical solution to the last equation is plotted in figure 5.3c, showing that the (second order) transition from the disconnected ($z < 1$) to the connected ($z > 1$) phase occurs suddenly at the critical point z_c.

The importance of this phenomenon is obvious in terms of the collective properties that arise at the critical point: systems-level communication becomes possible and information can flow

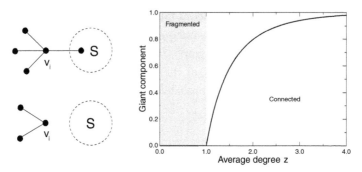

Figure 5.3. In order to estimate the value of the giant component S for the ER graph, we consider the probability of a given node v_i to belong (a) or not (b) to the giant component. In (c) we show the fraction S of the graph belonging to the giant component as the average degree z is changed. A phase transition occurs at $z_c = 1$, where the fraction of nodes included in the giant component rapidly increases.

from the units to the whole system and back. Moreover, the transition occurs suddenly and implies an innovation. No less important, it takes place at a low cost in terms of the number of required links. Since the only requirement for reaching whole communication is to have a given (small) number of links per node, once the threshold is reached, order can emerge *for free*. In this context, an additional result from the ER analysis concerns global connectedness. Specifically, when $p \geq \ln N/N$ (i.e., for $z > \ln N$), almost any graph in $G_{N,p}^{ER}$ is totally connected.

5.4 Real Networks Are Heterogeneous

The present example is important in providing a shortcut to the percolation problem already discussed in chapter 4. By ignoring lattice effects (and thus spatial correlations), we obtain an extremely simple model of a system formed by a large number of connected units. But we can also ask ourselves how general the

ER approach can be. Are real networks at all like this? The answer is essentially negative. Real networks have been found to diverge from this homogeneous picture in a number of ways (Albert and Barabási 2002; Dogorovtsev and Mendes 2003; Newman 2003b).

Perhaps the most interesting feature exhibited by most real networks, from the genome to the Internet, is that they are highly heterogeneous: they can be described by a degree distribution of the form $P(k) = (1/Z)k^{-\gamma}\phi(k/\xi)$ where $\phi(k/\xi)$ introduces a cut-off at some characteristic scale ξ and Z is a normalization constant.[3] On a log-log display, these distributions will show, for high cut-offs, a straight line behavior. As the value of ξ grows, fat tails develop.

Three main classes can be defined (Amaral et al. 2000): (a) When ξ is very small, $P(k) \sim \phi(k/\xi)$ and thus the link distribution is homogeneous. Typically, this would correspond to exponential or Gaussian distributions; (b) as ξ grows, a power law with a sharp cut-off is obtained; (c) for large ξ, scale-free nets are observed. The last two cases have been shown to be widespread and their topological properties have immediate consequences for network robustness and fragility. In particular, since a scale-free shape implies that most elements will have just one or a few links, the removal of such nodes will have little effect on the network (particularly in relation to its connectivity). In this context, these webs are expected to be robust against random failures. However, the presence of a few, highly connected hubs creates some undesirable effects: if they are removed (or fail to properly operate) the whole system can experience a collapse.

---

[3] This normalization is defined as $Z = \sum_{k=1}^{K} k^{-\gamma}\phi\left(\frac{k}{\xi}\right)$ where K is the maximum number of links that a node displays within the network.

6

LIFE ORIGINS

6.1 Prebiotic Evolution

How life originated on our planet is considered one of the most fundamental questions of science (Maynard Smith and Szathmáry 1999; Dyson 1999, Kauffman 1993). The emergence of living structures seems inconsistent with the expected patterns of decay and increasing disorder displayed by nonliving matter. And yet, somehow self-replicating structures appeared and evolved into life forms of increasing complexity. Although cells have become the basic units of living organization, the initial steps toward life likely involved the emergence of some classes of complex chemical reactions leading to selection for efficient replicators not necessarily associated with closed membranes (Maynard Smith and Szathmáry 1999; Schuster 1996; Ricardo and Szostak 2009).

Under prebiotic conditions, nonliving matter was composed of a rich soup of diverse molecules, some of them generated on the earth's surface and others—as suggested by the Catalan astrochemist Joan Oró—likely coming from comets falling on our planet (Oró 1961a). Such molecules included the building blocks of life, for instance, amino acids, nucleotides, lipids, and sugars, most of which have been shown to spontaneously emerge under laboratory conditions out of water, methane, hydrogen, and other nonbiological molecules (Miller 1953, Oró 1961b). But such precursors are not enough for life to emerge: in order to go beyond the soup, some chemical reactions need to be amplified so that the abundance of certain special molecules increases and allows selection processes to be at play (Eigen 1999). If such amplification occurs, growth of a subset of chemical species would take place. And growth is clearly a first ingredient necessary for reproduction. Afterward, Darwinian selection can come into play (Szathmáry 2006).

Growth is easily achieved through autocatalysis when a given component A can make a copy of itself by using available monomers E, that is,

$$A + E \xrightarrow{\mu} 2A + W \tag{6.1}$$

being W waste material and μ the reaction rate. This system can be described using a Malthusian growth process, that is, if $x = [A]$, we have $dx/dt = \mu x$ which gives (see chapter 2) an exponential growth dynamics, which is to say, $x(t) = x_0 e^{\mu t}$ which is obviously valid only for some restricted conditions, since sooner or later spatial constraints will slow the dynamics (such as in a logistic system).

6.2 Replicators and Reproducers

Once autocatalysis is at work, the question is what type of complex chemical structures can be sustained and what are the minimal requirements for such autocatalytic systems to emerge. Cooperation has been suggested as an essential step toward the emergence of complex, self-organized chemical systems (Eigen 1971). In particular, we can consider the effects of having cooperation among molecules in order to generate new copies, namely reactions such as

$$2A + E \xrightarrow{\mu} 3A + W \tag{6.2}$$

For this system, it is not difficult to see that the corresponding equation is $dx/dt = \mu x^2$ and the solution now takes the form

$$x(t) = \frac{1}{1/x_0 - \mu t} \qquad (6.3)$$

It can be easily shown that this solution has a well-defined asymptote at a finite time $t_c = 1/\mu x_0$ where a divergence in the concentration is observed. This means that after a well-defined time interval, an infinite concentration will be reached. This type of dynamics is known as *hyperbolic* and as we can see is dependent upon the initial state. Through competition among several chemical species (or replicators), the ones with higher initial populations would be more likely to win the race.

Here we want to analyze the properties displayed by a system where replicators (either molecules or cells) show both Malthusian and cooperative dynamics, following the work by Ferreira and Fontanari (2002).

6.3 Autocatalysis and Cooperation

We will consider a simple situation (figure 6.1) where the replicators A change their concentrations through autocatalysis, cooperation, and decay:

$$A \xrightarrow{s} 2A \qquad 2A \xrightarrow{\mu} 3A \qquad A \xrightarrow{1} 0 \qquad (6.4)$$

where the last one occurs at a unit rate. For simplicity, we do not include the precursors or waste materials here (although both are always present) since they would not affect our calculations provided that precursors are abundant.

In the following, we will assume that reactions occur in a system with limited space and that the available monomers are present in high numbers. For convenience we also fix $[E] = 1$. If we consider a normalized concentration $0 \le x \le 1$ using the

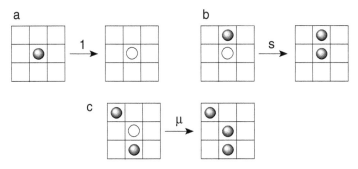

Figure 6.1. A simple model of replicating molecules considers three basic reactions, here indicating as happening on a two-dimensional lattice. Reactions include: (a) decay, where a given particle (gray sphere) spontaneously degrades (empty circle); (b) autocatalysis, where a given particle occupies an empty spot and (c) cooperative replication, where the presence of a molecule in the neighborhood helps the replication of another molecule.

previous set of reactions, we have a mean field model for $x = [A]$

$$\frac{dx}{dt} = -x + sx(1 - x) + \mu x^2(1 - x) \qquad (6.5)$$

where the three terms in the right-hand side of the equation correspond to the three mechanisms indicated in the previous figure.

An obvious stationary state corresponds to the extinction configuration (i.e., $x^* = 0$). The stability of this state is obtained from

$$\lambda_\mu = \frac{\partial \dot{x}}{\partial x} = -1 + s(1 - 2x) + \mu x(2 - 3x) \qquad (6.6)$$

which gives $\lambda_\mu(0) = s - 1$, and thus the extinction state will be stable if $s < 1$.

Two additional fixed points can be present, and correspond to the solutions of the quadratic equation $-1 + s(1 - x) + \mu x(1 - x) = 0$. This gives:

$$x^*_\pm = \frac{1}{2\mu} \left[\mu - s \pm \sqrt{(\mu + s)^2 - 4\mu} \right] \qquad (6.7)$$

By analyzing this expression, it is not difficult to find the conditions for the presence or absence of the nontrivial fixed points. First, real solutions will exist provided that

$$(\mu + s)^2 > 4\mu \tag{6.8}$$

and these solutions will be positive (and thus meaningful) provided that the inequality $\mu - s > 0$ holds, and thus an additional constraint to consider is $\mu > s$. From the condition of real roots, we can find the critical boundary:

$$s(\mu) = -\mu + 2\sqrt{\mu} \tag{6.9}$$

These conditions allow defining three phases in the (μ, s) parameter space, indicated in figure 6.2a. These correspond to: (a) an active phase where molecules are always present (upper domain, light gray), (b) an empty phase where the population goes extinct, and (c) a bistable phase (dark gray) where two alternate states are possible. The last two phases are separated by the critical curve $s(\mu)$ defining the boundary for a first-order transition.

The bistable character of our model is clearly observed in the bifurcation diagram shown in figure 6.2b, where we plot the fixed points against the cooperative replication rate μ for a fixed s value (here $s = 0.75$). As predicted by the stability analysis, the extinction point always appears stable: small population deviations will fall off to zero. A marked change occurs at a critical rate, which can be obtained from our previous equations for each s value and is simply $\mu_c = 2 - s + 2\sqrt{1 - s}$.

The presence of a first-order transition has two important implications for our problem. First, either an efficient autocatalysis or a combination of autocatalytic and cooperative effects needs to be present in order to sustain a stable population. Second, a minimal initial population must be stabilized in order to avoid falling into the extinct state. We can apreciate this situation by

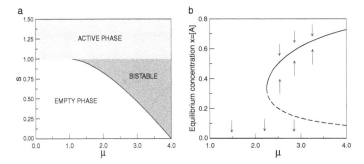

Figure 6.2. Phases in the model of molecular replicators. In (a) the three basic phases are shown. The black line separating the empty (extinction) phase from the bistable phase is given by equation 6.9. In (b) an example of the bifurcation diagram is shown for $s = 0.75$. For this parameter value, we have a bifurcation at $\mu_c = 0.25$.

using the potential function associated with our system, namely:

$$\Phi_\mu(x) = (1 - s)\frac{x^2}{2} + (s - \mu)\frac{x^3}{3} + \mu\frac{x^4}{4} \qquad (6.10)$$

which is plotted in figure 6.3. Here we use $s = 0.75$ and three different values of μ. The minima defining the alive phase coexist with an alternative minimum where extinction is also an alternative possibility. When $\mu < \mu_c = 2.25$, a unique minimum is observable, associated with the extinction scenario (or *dead* phase), whereas for $\mu > \mu_c$, we will observe two minima, being the *alive* fixed point placed in a deeper valley. Moreover, since fluctuations are expected to be involved in any of these prebiotic scenarios, noise could push the system from the alive to the dead state, thus making the transition less robust than predicted from our deterministic framework.

6.4 Prebiotic Reaction Networks

The previous results, as well as others suggested by theoretical models (see next chapter) indicate that early life might have

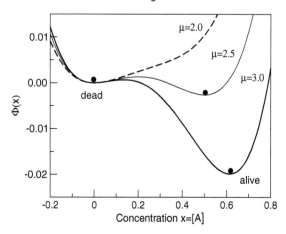

Figure 6.3. Stable states in the replicator model describe two basic types of prebiotic organization, namely alive and dead. The first corresponds to a stable, nonzero concentration of A molecules, whereas the second deals with extinction. If the reactivity is too low, decay will win over synthesis, but even when the reaction rates are appropriate to sustain a stable population, a minimal initial number is required. The potential function $\Phi_\mu(x)$ provides a good picture of the problem. Here three examples are shown (see text).

faced some serious obstacles. The need for a minimal amount of molecules allowing the system to reach a high concentration of self-sustaining molecules might indicate that the jump from nonliving to living matter was difficult. However, it is known that life originated very early on our planet, almost as soon as its surface cooled enough to allow a stable chemistry to exist.

In trying to understand the possible origins of life, some theories have used the idea that molecular species can interact in diverse ways and create cooperative structures (Eigen and Schuster 1979; Farmer et al. 1986; Kauffman 1993). The underlying idea is that a molecular soup receiving external energy inputs and displaying enough chemical diversity will eventually generate a self-sustained autocatalytic set. This argument is based on two key assumptions. First, available molecules must exhibit catalytic properties, that is, they must actively contribute to allow for new reactions to occur. The second is that, as the molecular size of the polymers grows, the number of chains that can be created explodes. In his early contribution to this problem, Kauffman derived a percolation condition for these autocatalytic sets to emerge (Kauffman 1993). Once such a percolation threshold was reached, complex networks would spontaneously form and persist. These networks would also sustain a high diversity of chemical molecules. Such diversity can be properly analyzed by using the phase transition approach (Jain and Krishna 2001; Hanel et al. 2005; see also Szathmáry 2006).

7

VIRUS DYNAMICS

7.1 Molecular Parasites

Life is rich in variety. Over hundreds of millions of years, life forms have been emerging in all habitats around the world. At all scales, from nanometers to meters, the success of life is obvious and overwhelming. Complexity has been achieved by means of different sources of change and innovation. These include multicellularity, cooperation, and language. But there is a dark side common to all these life forms: the presence of parasites. Parasitic entities seem a universal outcome of life. No known living species seems to be free from at least one type of parasite. The most commons forms of these entities are viruses, which are molecular machines that exploit cellular complexity to replicate themselves. Their genomes are typically small, too small in any case to allow them autonomy, but large enough to allow them to enter their hosts and use the host's molecular machinery for their own purposes.

Viruses have a large diversity of structures and genomes (figure 7.1). In terms of genetic material, there are DNA and RNA viruses with single-stranded, double-stranded, and fragmented genomes. RNA viruses offer a unique opportunity for exploring long-term evolution under controlled conditions due

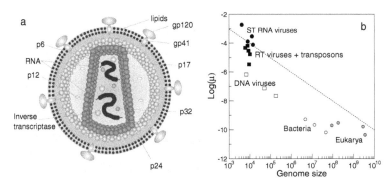

Figure 7.1. In (a) we show an example of RNA virus (HIV) where different proteins and other components are indicated. RNA viruses display high mutation rates, which scale as the inverse of their genome length. This pattern of decay is roughly followed by other organisms, as shown in (b) where we plot the logarithm of mutation rate μ against the logarithm of genome size. The dotted line is the predicted inverse relation among both quantities (see text). Figure (a) has been adapted from commons.wikimedia.org/wiki/VIHsanslibel.png.

to their high mutation rates. They are the most important class of intracellular parasites and are responsible for a wide range of diseases (Domingo 2006). They are extremely adaptable to changing conditions, particularly to selection barriers imposed by immune responses. The immune system is able to identify pieces of the infecting virus, eventually wiping out the original strain. However, if the virus changes fast enough, it can escape from immune recognition. Since high mutation rates allow viruses to escape, one might conclude that the higher the mutation, the better. Actually, DNA-based life forms are known to support sophisticated mechanisms for error correction, whereas RNA viruses lack such molecular machinery.

The enzymes allowing RNA viral genomes to replicate are in fact rather inaccurate. What are the limits to such error rates? We know that mutations have deleterious effects: a single change in a nucleotide can easily make the new sequence unable to replicate at all. Does such a conflict between change and conservation exhibit some optimal trade-off? In the early 1970s, a theoretical model of replicating and mutating populations revealed that there must be an error threshold beyond which information cannot be preserved. It was suggested that due to mutation, these populations or *quasispecies* would be extremely heterogenous (Eigen 1971; Mas et al. 2010). The quasispecies structure has numerous implications for the biology and associated pathology of RNA viruses, the most relevant of which is that the heterogeneous population structure is a reservoir of variants with potentially useful phenotypes in the face of environmental change.

Mutation rates cannot reach arbitrary values: there is a threshold to mutational change beyond which no selection is possible. This threshold has been dubbed the *error threshold* (Schuster 1994; Domingo 2006). Roughly, it was predicted that the mutation rate μ_n per nucleotide and replication round should be $\mu_n^c \approx 1/v$, with v the sequence length. The Eigen-Schuster theory predicts that, beyond μ_c, genomic information is lost as the population enters into a drift phase (Eigen 1971; Schuster 1994). Such prediction received strong support from the first estimations of viral mutation rates for different RNA viruses. Actually, it turns out that RNA viruses live right at the error threshold. In this chapter we will explain the origins of this behavior within the context of phase transitions. As will be shown, the boundary separating information preservation and randomness is a sharp one.

7.2 Exploring the Hypercube

Let us first consider a simulation experiment that provides some intuitions about the error catastrophe and its origins. Modeling virus populations would require considering cells, how they recognize their host, how they enter the cell, how their molecules are copied, how they exchange information with the host cell, and how the new, mutated copies, leave the cell. However, we might not want to know all the details of their life cycle. Instead, we ascertain the limits of the mutability of a given RNA virus and how these relate to genome size. As usual, we need to define a toy model, where a simple representation of the virus genome is taken into account and a complete abstraction of the underlying host properties is considered. In such description, a population of viruses is considered. Viruses will be strings of v units and will replicate with a given accuracy.

In our approximation, a digitial virus is defined as a string $\mathbf{S}_k$ with $k = 1, \ldots, N$, being N the total population size. Each string represents a small genome of size v, that is,

$$\mathbf{S}_i = (S_i^1, S_i^2, \ldots, S_i^v) \qquad (7.1)$$

where each element is a Boolean variable, $S_k^i \in \{0, 1\}$. Formally, we can think of each possible string as a vertex $\mathbf{S}_k \in \mathcal{H}^v$ of a v-dimensional hypercube (see figure 7.2). The hypercube is an extremely large object. The examples shown in figure 7.2 illustrate how we can construct such a landscape under our simplistic assumptions. But for a real virus, even the smallest ones, v is a large number, ranging from $v \approx 400$ for plant viroids (see Gago et al. 2009 and references therein) to $v \sim 10^4$ for most well-known RNA viruses responsible for infectious diseases.[1]

Let us allow the system to change and evolve. Our model starts with an initial population of master sequences and repeats, at each generation, N times, the following set of rules:

1. We take a string $\mathbf{S}_i$ at random from the population and replicate it with probability $f(\mathbf{S}_i)$. Here two replication probabilities are also defined, one for the master

---

[1] This means that the number of nodes in $\mathcal{H}^v$ will be hyperastronomic, much larger in fact than the number of atoms in our universe.

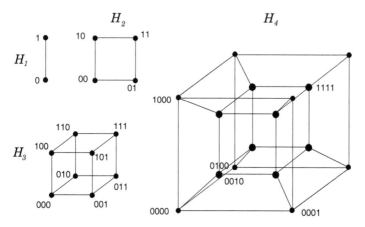

Figure 7.2. Quasispecies live in a hypercube. In its simplest form, strings are represented as chains of bits of fixed length v. Here the hypercubes $\mathcal{H}^v$ for $v = 1, 2, 3, 4$ bits are shown. Populations flow between the vertices of each hypercube, being two vertices connected through single point mutations.

sequence (where $S_i^k = 1$ for all $k = 1, \ldots, v$) and the other for the rest of the strings, to be indicated as f_1 and f_2, respectively. Hereafter we indicate the master sequence as $\xi = (1, 1, \ldots, 1)$ and assume that $f_1 \geq f_2$.

2. Replication takes place by replacing one of the strings in the population (also chosen at random), say $\mathbf{S}_j \neq \mathbf{S}_i$ by a copy of $\mathbf{S}_i$. The copy mechanism takes place with some errors. Such errors take place with some probability (or *mutation rate* μ_b) per bit and replication cycle respectively. In this way, each bit S_i^k can be replaced by $1 - S_i^k$ with probability μ_b.

The result of our simulations is shown in figure 7.3, where we plot three examples of the time evolution of the number of strings, starting from our special initial condition. A different

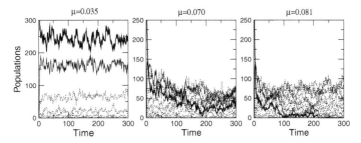

Figure 7.3. Time evolution of the bit string quasispecies model described in the text. Here we use a population of $N = 500$ strings, with $f_1 = 1$ and $f_2 = 0.25$. For small mutation rates (left plot) the master sequence (thick line) dominates the population, followed by the populations of strings one-bit distant in sequence space. These are lumped together in a single population, here indicated as a thin continuous line. Other populations with strings differing two, three, or more bits from the master are also indicated using dotted lines. As we increase μ_b the master sequence becomes less and less frequent: in the middle plot we can see that it falls below other populations, but still persists. For higher mutation rates, the master sequence disappears (right panel).

representation is provided by the stationary values of each population (figure 7.4). For low mutation rates, the master sequence remains high (thick line) whereas other mutant sequences have lower populations. For simplicity, we aggregate all strings differing one, two, or more bits from the master string. This is typically measured by means of the Hamming distance d_H:

$$d_H(\mathbf{S}_i, \mathbf{S}_j) = \sum_{k=1}^{\nu} |S_i^k - S_j^k| \tag{7.2}$$

As μ_b grows, we can observe that the master sequence starts to fall below that of other populations, eventually disappearing beyond some critical value.

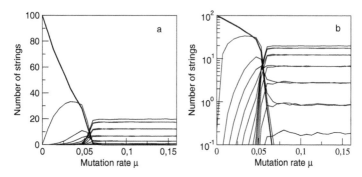

Figure 7.4. The error catastrophe: using a bit string model, with $N = 100$ strings of size $v = 16$, starting from an initial condition involving only master sequences. After $2000 \times N$ replications with mutation, we plot the number of master sequences (thick line) as well as the number of strings with a Hamming distance $d_H = 1, 2, \ldots, 16$ from the master. Although the master dominates at low mutation rates, a sharp change takes place at a critical mutation rate, where the master sequence vanishes as the quasispecies starts wandering through sequence space at random. Here linear (a) and linear-logarithmic (b) plots have been used.

7.3 Swetina-Schuster Quasispecies Model

In its simplest form, we can consider a reduced system of equations defining a population formed by two basic groups: the master sequence x_1 and the other sequences, which we assume to be grouped into an "average" sequence with population x_2 (Schuster 1994). Let us also assume (as a first approximation) that mutations occur from the master to the second compartment but not in the reverse sense. The enormous size of the sequence space $\mathcal{H}^v$ makes this assumption a good first approximation. Each time a master sequence replicates, there are v possible neighboring mutants. If we consider a real virus, this means a very large number of ways of leaving the master node and diffusing through sequence space. To move out from a given node

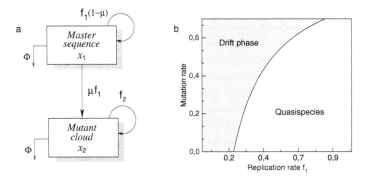

Figure 7.5. Phase transitions in the Swetina-Schuster model. In (a) we summarize the basic structure of the two-compartment model. In (b) the two phases are shown, as obtained from the mean field approximation. Here we represent mutation rate per genome (i.e., the probability that some unit in the genome gets mutated) against replication rate of the master sequence. The gray region indicates the disordered (drift) phase, where the virus population wanders randomly through sequence space. The white domain indicates the presence of a stable cloud of strings around the master sequence.

is thus easy, whereas going back is unlikely. The basic scheme is summarized in figure 7.5a. The two populations are linked through mutation from x_1 to x_2 (but not back). Additionally, each population replicates at well-defined rates, and an outflow rate Φ is also included in order to maintain populations at a finite size.

The model is given by the following set of coupled differential equations:

$$\frac{dx_1}{dt} = f_1(1 - \mu)x_1 - x_1\Phi(x_1, x_2) \qquad (7.3)$$

$$\frac{dx_2}{dt} = f_1\mu x_1 + f_2 x_2 - x_2\Phi(x_1, x_2) \qquad (7.4)$$

where μ is the mutation rate, f_1 is the master sequence replication rate, and f_2 is the replication rate for the other strings. The term $\Phi(x_1, x_2)$ will allow selection to be effective. This is done by assuming a constant population (CP) constraint, namely that $x_1 + x_2$ is kept constant. This means that

$$\frac{d(x_1 + x_2)}{dt} = \frac{dx_1}{dt} + \frac{dx_2}{dt} = 0 \qquad (7.5)$$

which allows defining the outflow term as:

$$\Phi(x_1, x_2) = \frac{f_1 x_1 + f_2 x_2}{x_1 + x_2} \qquad (7.6)$$

Since the relative concentrations of x_1 and x_2 are given by

$$P_1 = \frac{x_1}{x_1 + x_2} \qquad P_2 = \frac{x_2}{x_1 + x_2} \qquad (7.7)$$

we have in fact demonstrated that the outflow is nothing but the average replication rate:

$$\Phi(x_1, x_2) = P_1 f_1 + P_2 f_2 = \langle f \rangle \qquad (7.8)$$

For simplicity, we fix $x_1 + x_2 = 1$ in such a way that the x_i's are already probabilities. Using this constraint, the previous model can be reduced to a one-dimensional system (using $x_2 = 1 - x_1$) and this leads to:

$$\frac{dx_1}{dt} = f_1 x_1 [\xi_1 - \xi_2 x_1] \qquad (7.9)$$

where $\xi_1 = 1 - \mu - f_2/f_1$ and $\xi_2 = 1 - f_2/f_1$. Here the fixed points are $x_0^* = 0$, $x_1^* = \xi_1/\xi_2$, and thus the nontrivial fixed point, representing a nonzero master sequence population, will be:

$$x_1^* = 1 - \frac{\mu f_1}{f_1 - f_2} \qquad (7.10)$$

The linear stability analysis shows that $x_0^* = 0$ is stable (and the master sequence disappears) if

$$\mu > \mu_c = 1 - \frac{f_2}{f_1} \qquad (7.11)$$

whereas a stable cloud of mutants coexisting with the master sequence will be stable for $\mu < \mu_c$. The critical mutation rate separates the two basic phases. In figure 7.5b we show these phases, separated by the critical line, using a fixed f_2 value (here we use $f_2 = 0.25$). Once such a boundary is crossed, we shift from one type of qualitative dynamics to the other. The standard-error threshold condition is associated with an increase in mutation rate. Increased μ values crossing the critical line drive the master sequence into extinction.

7.4 Critical Genome Size

A different form of the previous critical condition can be derived by explicitly using the ν parameter. This derivation will allow us to recover the inverse relation between mutation rates and genome size. In order to obtain such a result, let us first write μ as a function of ν and μ_b (mutation rate per bit). It is not difficult to see that they are related through:

$$\mu = 1 - (1 - \mu_b)^{\nu} \tag{7.12}$$

where we have used the probability of no mutation of a given unit, $1 - \mu_b$, and considered the probability $p = (1 - \mu_b)^{\nu}$ that none of the nucleotides is mutated. The difference $1 - p$ is just the probability that some unit does mutate. Since μ_b is small, we can use the approximation[2]:

$$\mu \approx 1 - e^{-\mu_b \nu} \tag{7.13}$$

It is not difficult to show that the previous critical condition is now written, for the mutation rate per unit, as

$$\mu_b^c = \frac{\alpha}{\nu} \tag{7.14}$$

---

[2] Here we make use of the first two terms of the Taylor expansion of the exponential function, i.e., $e^x \approx 1 + x + x^2/2! \ldots$.

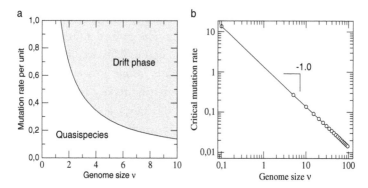

Figure 7.6. The two phases displayed by the Swetina-Schuster model using the mutation rate per unit and round of replication against gemome size. The critical line separating the two phases in (a) gives the mutation-genome size observed in real RNA viruses, shown in (b) in log-log scale. The resulting power law shows that the critical mutation scales as the inverse of genome length.

where $\alpha = -\ln(f_2/f_1) > 0$ is a constant. The last expression actually corresponds to the observed inverse decay of mutation rates in RNA viruses as an inverse of their genome size. The resulting parameter space is also shown in figure 7.6a. Moreover, if we plot the critical line $\mu_b^c(\nu)$ in log-log scale (here we use $f_1 = 1.0$), a well-defined power law is recovered (figure 7.6b). An alternate derivation of this condition using a simple percolation-like argument can be obtained based on branching processes (Demetrius et al. 1985; Hofbauer and Sigmund 1991).

7.5 Beyond the Threshold

Increased mutagenesis beyond the error catastrophe can destroy the virus, since beyond the threshold no Darwinian selection is at work (Schuster 1994). Moreover, considering the fact that many mutants will be lethal (at any mutation rate), several critical thresholds can be obtained combining lethality and catastrophe

(Bull et al. 2005). The exceptionally high mutation rate in RNA viruses is illustrated by the finding that most HIV virions in blood appear to be nonviable. Effective experimental strategies have shown that the error threshold can actually be exploited in antiviral therapy (Loeb et al. 2003). Within the context of HIV treatment, using promutagenic nucleoside analogs, viral replication of HIV has been shown to be abolished in vitro (Loeb et al. 2003). Moreover, it has been suggested that genetic instability in cancer cells might lead to an error catastrophe: mutations and chromosomal changes can increase up to some limit (Solé 2003; Solé and Deisboeck 2004; Gatenby and Frieden 2002; see also chapter 11).

Our previous models consider a simplistic fitness landscape where one given string has a large fitness in relation to all other possible strings. Such situation is of course an oversimplification. Fitness landscapes are in general much more complex and rich (Bull et al. 2005; Jain and Krug 2007), and other components such as space (Altmeyer and McCaskill 2001; Pastor-Satorras and Solé 2001) and recombination (Boerjlist et al. 1996) can also play an important role.

Many developments of the quasispecies model have been considered since its original formulation, considering the effects of multiple peaks, correlations, and time-dependent changes. In this context, an important finding was the presence of neutrality in RNA landscapes (van Nimwegen et al. 1999; Wilke 2001a, b). In a nutshell, many mutations are selectively neutral (Kimura 1983) and it has also been found that chains of single mutations allow connections between neutral genotypes in sequence space (our hypercube) in such a way that these *neutral networks* percolate through large domains of such space (Schuster et al. 1994; van Nimweggen et al. 1999; van Nimweggen and Crutchfield 2000; Reydis et al. 2001). Neutral networks allow populations to display high robustness against mutations and have some unexpected implications. One of them is that, under high mutation rates, selection can sometimes favor slower—but more robust—genomes. Such a *survival of the flattest* effect has been found both in simulated (Wilke et al. 2001; Sardanyes et al. 2008) and real (Codoñer et al. 2006; Sanjuan et al. 2007) systems.

8

CELL STRUCTURE

8.1 The Architecture of Cells

Cells are the basic units of life. They have often been compared to machines, involving many components in interaction. These components, particularly proteins, define a set of functional structures able to cope with all cell requirements, including the gathering and processing of matter, energy, and information (Harold 2001). Moreover, they also sustain the internal structural organization of cells: proteins are the basic units from which complex chains of polymers are formed. Some of these polymers allow cells to define their shapes. They form the so-called cytoskeleton (Alberts et al. 2004; Pollard 2003). This skeleton functions a scaffold, a positioning system, a distribution grid, and a force-generating apparatus used for locomotion (Karp 1999; Pollack 2001). Most of this molecular machinery is constantly changing, and its behavior lies somewhere between order and disorder. Understanding how they coordinately organize in space and time is a great challenge to experimental and theoretical approaches alike.

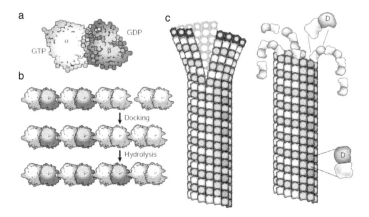

Figure 8.1. Microtubules are formed by tubulin dimers (a) involving two almost identical subunits, labeled α and β respectively. These dimers polymerize, eventually forming long, cylindrical structures. In (b–c) the dynamical patterns associated with growth and decay of the microtubule are shown (adapted from Howard and Hyman 2003). They grow by addition of GTP-bound tubuline, and shrink through hydrolysis. If hydrolysis dominates, the cap disappears and the structure rapidly dissociates.

Microtubules are one particularly important component of cell organization.[1] They are not essential only to cell architecture. They are highly dynamical, in that they are involved in several key cellular processes such as cell division, motion, and transport (Howard and Hyman 2003). They are an essential factor for cell stability, and actually some anticancer treatments involve molecular interactions between drugs and the proteins forming the microtubules. They can be properly described as cell railways, since they provide the physical substrate along which transport of cargo takes place. The basic molecular organization of these complex polymers is outlined in figure 8.1. The dynamical

---

[1] The other two are the actin filaments and the so-called intermediate filaments. The two involve a diverse range of biological polymers. All three groups interact through additional proteins.

properties of microtubule formation strongly differ from other self-assembling systems, such as ribosomes and viruses. The latter involve stable structures, whereas microtubules are in a constant state of change and display a broad range of lengths. As pointed out in (Gerhardt and Kirshner 1997) they seem to avoid equilibrium.

Microtubules are long, hollow tubular structures. They are assembled from dimeric building blocks, combining two very similar proteins (figure 8.1a), so-called α-tubulin and β-tubulin, respectively. Each microtubule is like a cylinder composed of thirteen parallel filaments having a well-defined polarity. The two ends of the tube are usually indicated as the $+$ and $-$ ends. The elongation of the microtubules takes place at the $+$ end. Polymerization and depolymerization take place all the time within cells, and this process displays many fluctuations. A given microtubule can grow steadily over a long period of time and then rapidly shrink, a behavior known as *dynamic instability*.

The driving force of the dynamic instability is the transformation of a small molecule, GTP, into another molecule GDP.[2] GTP binds to both tubulin subunits, but only those associated to the β can experience the $GTP \rightarrow GDP$ change, also known as the hydrolyzation of GTP. When the GTP is hydrolyzed, the chain becomes curved and thus the cylinder becomes less stable. This process is delayed and can take place at different speeds depending on local conditions. If the polymerization process is faster than the rate of $GTP \rightarrow GDP$ conversion, growth will occur. If not, the microtubule will shrink. The balance between these two basic processes defines two possible phases. A cell can tune these rates accordingly in order to favor growth or allow shape changes. Our goal here is to find the boundary separating these two phases using a minimal model of polymerization dynamics.

---

[2] GTP and GDP stand for guanidine tri- and di-phosphate, respectively.

8.2 States and Transitions

In order to define a model of microtubule dynamics that could be solved in terms of a simple, one-dimensional dynamical system, several drastic assumptions must be made. Here we follow the approach taken by Antal, Krapivsky, and Redner (AKR) to tackle this problem (Antal et al. 2007a, 2007b). First, we will consider our problem as an essentially one-dimensional one, where a microtubule would be a linear string. Moreover, instead of worrying about the dimeric nature of tubulin, we will simply consider each bead in the linear chain as a monomer. This decision seems reasonable since the relevant component affects just one of the tubulins (the β unit).

Using this view, we now concentrate on what is going on at the tip of the string. In particular, we can look at the end of the string and see what is located in the last position. We will indicate as filled and empty circles the GTP and GDP subunits, respectively. Following the AKR approach, we will indicate the two possible terminal states as follows:

$$| \cdots \bullet \rangle \tag{8.1}$$

for the GTP ending tip and

$$| \cdots \circ \rangle \tag{8.2}$$

for the GDP one. The dots simply indicate the existence of a linear chain, no matter its composition. Additionally, we will use $| \ldots \rangle$ to indicate any string, independent of how its tip is defined.

Several possible key transitions can be defined here, associated to how the tip of our microtubule changes. In the AKR model, the following four changes were considered.

1. A GTP end can incorporate a new GTP at a rate λ:

$$| \cdots \bullet \rangle \xrightarrow{\lambda} | \cdots \bullet \bullet \rangle \tag{8.3}$$

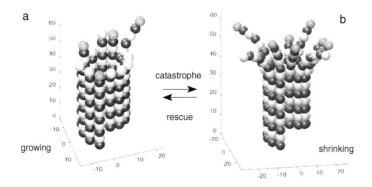

Figure 8.2. Microtubules are constantly changing. Two basic types of dynamical behaviors are at work: either they grow smoothly by adding new tubulin dimers (a) or rapidly shrink (b).

2. A GDP end can grow by incorporating a GTP and at a rate $p\lambda$:

$$| \cdots \circ \rangle \xrightarrow{p\lambda} | \cdots \circ \bullet \rangle \qquad (8.4)$$

3. Any intermediate (nonending) GTP unit can transform into a GDP end at a rate one:

$$| \cdots \bullet \cdots \rangle \xrightarrow{1} | \cdots \circ \cdots \rangle \qquad (8.5)$$

4. A GDP ending can be removed at a rate μ:

$$| \cdots \circ \rangle \xrightarrow{\mu} | \cdots \rangle \qquad (8.6)$$

This set of rules can be used as the basis for a mean field approximation, where the details of microtubule structure and fluctuations are ignored.

8.3 Dynamical Instability Model

Two key quantities need to be taken into account in order to derive the mean field model for this process. First, the fraction of

GDP ends, that is, the probability of having a ○ at the tip, which we will indicate as

$$n_0 \equiv P[|\cdots\circ\rangle] \tag{8.7}$$

By using the previous list of transitions, we can write a rate equation for the dynamics of n_0. This equation reads:

$$\frac{dn_0}{dt} = -p\lambda n_0 + (1 - n_0) - \mu \mathcal{N}_0 \tag{8.8}$$

where we indicate as

$$\mathcal{N}_0 \equiv |\cdots\bullet\circ\rangle \tag{8.9}$$

that is, the frequency of terminal tips ending in a GTP-GDP pair (in this order).

The second component in our analytic derivation is the speed $V(\lambda, \mu, p)$ at which the tip of the string is growing:

$$V(\lambda, \mu, p) = p\lambda n_0 + \lambda(1 - n_0) - \mu n_0 \tag{8.10}$$

The differential equation for n_0 cannot be easily solved, since the last term $\mathcal{N}_0$ would be difficult to determine. However, we can make a safe approximation by considering that the confirguration $|\cdots\bullet\circ\rangle$ is uncommon, since the GTP units not located at the end of the microtubule are rapidly transferred and thus $\mathcal{N}_0 \approx 0$. With this approximation, our previous dynamical equation now reads:

$$\frac{dn_0}{dt} = 1 - (1 + p\lambda)n_0 \tag{8.11}$$

which gives a single and stable fixed point:

$$n_0^* = \frac{1}{1 + p\lambda} \tag{8.12}$$

Now our interest is to find the critical conditions separating the two dynamical phases. Here the speed of growth equation will be very useful in defining a percolation-like condition: the tip will grow (or shrink) provided that $V(\lambda, \mu, p)$ is larger (smaller)

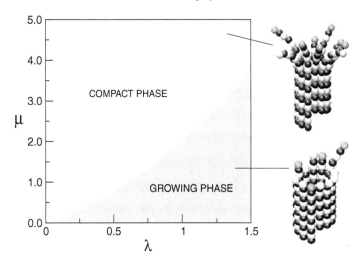

Figure 8.3. Phases of microtubule behavior from the mean field model. The two main types of behavior are described by the two areas separated by the critical curve $\mu_c = p\lambda(1 + \lambda)$ (here for $p = 1$).

than zero. The critical set of parameters is thus obtained from:

$$p\lambda n_0 + \lambda(1 - n_0) - \mu n_0 = 0 \qquad (8.13)$$

and now we can use the stationary state n_0^*, as previously calculated. The resulting critical curve is:

$$\mu_c = p\lambda(1 + \lambda) \qquad (8.14)$$

When $\mu > \mu_c$ we have the compact phase, whereas the grow phase is observed for $\mu < \mu_c$. These phases are shown in figure 8.3. Although the assumptions used above are rather strong, it can be shown that these results are a good approximation of the exact behavior (Antal et al. 2007a, 2007b).

8.4 Tensegrity

As mentioned at the beginning of this chapter, microtubules are one component of a large family of polymers involved in shaping cellular architecture. Most of them are highly dynamical, although different time scales are actually involved. They define a limited range of molecular mechanisms controlling cellular changes (Pollack 2001). The observation that a cell's cytoplasm behaves very much as a polymer makes it possible to explore its adaptive behavior and plasticity in terms of simple models.

Beyond microtubule dynamics, several relevant problems remain open within cell biology, particularly in relation to large-scale patterns of cell organization. One approach to this problem has been made by means of so-called tensegrity, a building principle originally described with in the context of architecture (Fuller 1961). Tensegrity systems can be defined as structures that stabilize their shape through continuous tension. The concept has been applied to cell biology (Ingber 1998; 2003) and allows connecting mechanical forces and biochemical processes, thus integrating global behavior and cell regulation. Given the key role played by dynamical transitions between different phases in this type of polymer (such as sol-gel changes, see Pollack 2001), phase transition models should play a key role in our understanding of cell architecture and its dynamical changes.

9

EPIDEMIC SPREADING

9.1 Spreading Diseases

Infectious diseases have played an important role throughout the entire history of human life (Garrett 1994; Diamond 1999). They have been a significant source of mortality and have served as strong selective forces. No example better summarizes the terrible effect of infectious diseases than the Black Plague, which caused terror throughout Europe during the years 1347–50. Upon its arrival from the East, it spread through Europe killing about a quarter of the population. The disease was terribly virulent and effective. About 80 percent of the people who were in contact with the disease died within two to three days.

Infectious diseases endemic in Europe such as measles, smallpox, influenza, and bubonic plague were transmitted by invading peoples and have decimated entire ethnic groups that had not been in contact with the diseases and so had evolved no immunity. These infectious diseases played an influential role in European conquests (Diamond 1999). For example, a smallpox epidemic devastated the Aztecs after the first attack by the Spaniards in 1520. It is believed that up to 95 percent of the pre-Columbian Native American population was killed by diseases introduced by Europeans. And similar outcomes took place in other parts of the globe.

Epidemic modeling has been a very active area of research for several decades. One of the simplest model approaches to this problem is the so-called contact process. In this simple framework, two types of populations are considered, namely infective (I) and susceptible (S) individuals. The first are infected and carry the pathogen with them while the second are not infected but can become so. The states of these individuals can change due to two basic processes: recovery ($I \rightarrow S$) and infection ($S \rightarrow I$), which occur with some fixedrates.[1] In figure 9.1a, the basic transitions are summarized for a one-dimensional case (a realistic situation requires at least two dimensions). While the recovery process is independent of the state of the neighbors, the infection process is not: the more neighbors are infected, the larger the probability of infection.

In figure 9.1b we show the results of a numerical simulation where we start from an initial condition with half (randomly chosen) individuals being infected (here indicated as black dots). Here we use $N = 100$ individuals. Three examples of the dynamical patterns are shown in the inset plots. We measure the probability of propagation by determining whether infected individuals remain after five hundred steps. By averaging over one hundred runs, we obtain a phase transition diagram displaying a sharp change. For our example, where the probability of recovery is fixed to $1/2$, the simulation reveals that epidemic spreading does not occur if the rate of infection is smaller than $1/2$ whereas a stable infected population is always observed otherwise. This implies the presence of a second-order (critical) phase transition and the existence of a threshold of infection rate that needs to be overcome in order to generate a stable infected population.[2] A percolation process is at work.

---

[1] Alternatively, the process can also model the occupation of a given space by a population under a stochostic process involving.

[2] This is actually an example of the absorbing phase transition mentioned at the end of chapter 1. The absorbing phase corresponds to the zero infection

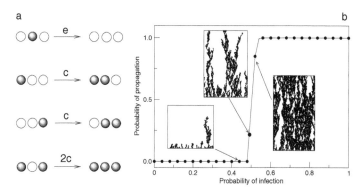

Figure 9.1. Epidemic spreading in a model based on a contact process. Here a one-dimensional lattice is considered, with each site occupied by an infective (gray ball) or a susceptible (empty ball). In (a) we show the four possible transitions that can take place: either an individual becomes susceptible (independent of the state of its neighbors) or a susceptible becomes infected. Three possible situations are allowed. If we measure the fraction of infectives after a long transient for different infection levels (b), the resulting curve displays a transition from no infectives to a sustained epidemic state (see text).

A theoretical treatment of the process including the spatial correlations is difficult (Hinrichsen 2000; Marro and Dickman 1997). If $\rho(\mathbf{x}, t)$ indicates the probability of infection at a given point $\mathbf{x} \in \mathcal{L}$, that is,

$$\rho(\mathbf{x}, t) = P[S_{\mathbf{x}}(t) = 1] \qquad (9.1)$$

then $\rho(\mathbf{x}, t)$ will evolve (according to the previous rules) following:

$$\frac{d\rho(\mathbf{x}, t)}{dt} = -\mu\rho(\mathbf{x}, t) + \frac{\alpha}{q} \sum_{<\mathbf{u}>}^{q} P[S_{\mathbf{x}}(t) = 0, S_{\mathbf{u}}(t) = 1] \quad (9.2)$$

---

phase. Once reached, there is no escape toward the second phase. Of course this is true under the hypothesis that the system is isolated and thus no infections can come from the external environment (which is clearly not the case in general).

where $<\mathbf{u}>$ indicates the sum over the set of q nearest neighbors. For one dimension, this set is simply $\{i - 1, i + 1\}$. The previous equation is exact, but its computation would require knowledge of the probabilities associated with the interactions between nearest sites. As will be shown below, a qualitative understanding of the problem is easily achieved from a simple mean field argument.

9.2 SIS Model

Here we define a minimal model of epidemic spreading, taking place between two populations, as previously defined for the contact process (see Murray 1990 and references therein). For simplicity, we will also indicate their population sizes as S and I.

The infection dynamics will include two basic processes: (a) new infections occur provided that infected individuals contact healthy ones. The rate at which infection takes place will be indicated as μ. On the other hand, (b) infected individuals can recover at a rate α, becoming healthy again. The two components of the process can be summarized in terms of two chemical-like reactions:

$$I + S \xrightarrow{\mu IS} 2I \qquad (9.3)$$

for the infection event and

$$I \xrightarrow{\alpha I} S \qquad (9.4)$$

for the recovery process. Additionally, let us assume that the total population size N remains constant, that is, $S + I = N$. As a result, we can write down the equation for the number of infectives as

$$\frac{dI}{dt} = \mu IS - \alpha I = \mu I(N - I) - \alpha I \qquad (9.5)$$

As we can see, we have two conflicting components defining the infection dynamics. One is a density-dependent term where the

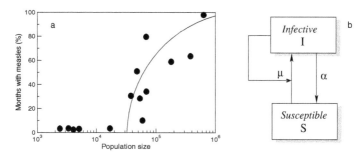

Figure 9.2. Epidemic spreading in the real world. In (a) we show
the number of months with measles as a function of population
size. Note the presence of a sharp change at intermediate population
values. A simple model of epidemic spreading (b) considers two sub-
populations composed of infected and healthy individuals. Although
infected ones recover with some rate α, newly infected ones require
interaction between both groups.

two groups need to interact, whereas the second is the linear
decay of infected individuals, as they recover from infection. This
equation has two equilibrium points, namely $I^* = 0$, describing
the all-healthy population state and a second one

$$I^* = N - \frac{\alpha}{\mu} \qquad (9.6)$$

describing a population with a finite number of infected individuals. It is easy to see that the stability of these states is determined
by the presence of a critical condition defined by the so-called
basic reproductive rate of the pathogen:

$$R_0 = \frac{\mu N}{\alpha} \qquad (9.7)$$

Epidemic spreading will occur if $R_0 > 1$, and decay will take
place instead for $R_0 < 1$. It is interesting to note that R_0 involves
several components, including the infectivity of the pathogen μ
but also the population size N. This implies that as population

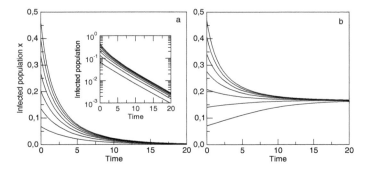

Figure 9.3. Time series of epidemic spreading. Here we use $\alpha = 1.0$ and (a) $\mu = 0.8$ (subcritical) and (b) $\mu = 1.2$ (supercritical). The inset in figure (a) is a linear-log display of the trajectories at the subcritical phase, showing a linear decay that corresponds to an exponential behavior.

size increases so do the chances for a pathogen to spread. This is actually a well-known observation, and an example is displayed in figure 9.2a for measles. Here we plot a surrogate of the previous order parameter using the (average) lifespan of measles outbreaks in different populations. Below some threshold, no effective spreading occurs, whereas after $N \approx 30000$, spreading seems guaranteed.

In order further analyze our model, we will use a normalization of population sizes, that is, $x = I/N$ and $y = S/N$ and thus using $y = 1 - x$, the previous equation reduces to:

$$\frac{dx}{dt} = \mu x(1 - x) - \alpha x \tag{9.8}$$

and the equilibrium points now read $x_0^* = 0$ and

$$x_1^* = 1 - \frac{\alpha}{\mu} \tag{9.9}$$

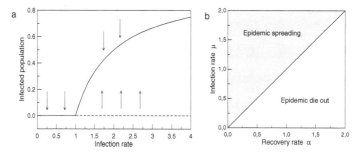

Figure 9.4. Phase transitions in the epidemic spreading model. In (a) the bifurcation diagram for the mean field model is shown, using $\alpha = 1.0$, giving $\mu_c = 1.0$. In (b) the two phases in parameter space (α, μ) are shown.

respectively. The epidemic will die out if

$$\lambda_\mu(0) = \mu - \alpha < 0 \qquad (9.10)$$

that is, if $\alpha > \mu$. The opposite case, defined for $\alpha < \mu$ corresponds to a stable epidemic spreading. The two points exchange their stability. In summary, as shown by the bifurcation diagram in figure 9.4a, the critical point $\mu_c = \alpha$ separates the two phases of this model: (1) the subcritical phase, where the epidemics dies out and (2) the supercritical phase, where the epidemics self-maintains. The corresponding parameter space is also displayed in figure 9.4b. The results essentially tell us that the balance between infection and recovery define the limits separating the two possible qualitative outcomes of the model.

9.3 Vaccination Thresholds

The presence of a critical threshold in epidemic dynamics implies that tipping points exist. Once a pathogen is able to cross the boundary, epidemic spreading will occur. But contention based on vaccination strategies can also exploit this threshold behavior in efficient ways. Consider our population, but now imagine that

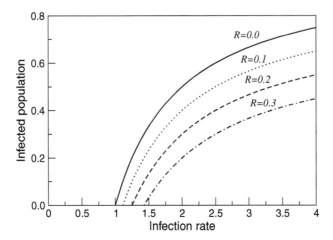

Figure 9.5. Effects of vaccination on epidemic spreading. Here different levels of vaccinated individuals R are introduced and compared with the predicted curve for nonvaccinated population (continuous line).

a fraction R of its individuals has been vaccinated and thus cannot be infected. Assuming normalized population size ($N = 1$), the new equation will read:

$$\frac{dx}{dt} = \mu x(1 - R - x) - \alpha x \tag{9.11}$$

with $0 \le R \le 1$. The reader can easily check that the new steady states for the infection phase are now

$$x_1^* = 1 - R - \frac{\alpha}{\mu} \tag{9.12}$$

which are stable if

$$\mu > \mu_c(R) = \frac{\alpha}{1 - R} \tag{9.13}$$

The resulting bifurcation curves are displayed in figure 9.5, where we can see that the eradication threshold shifts to higher values as the R parameter is increased. An alternate form for this threshold (as the reader can check) can be written as:

$$R > 1 - \frac{1}{R_0} \qquad (9.14)$$

(see Anderson and May 1991). The target of any eradication policy requires reducing R_0 to below one. Available data allows estimating this parameter. For smallpox, for example, we have $R_0 \approx 3 - 5$, and 60–70 percent of the population must be vaccinated. However, measles has a much higher basic reproductive rate of $R_0 \approx 13$, which requires a much larger vaccinated population (around 20 percent or $R \approx 0.92$). It is worth noting that smallpox has been eradicated from the world, and that measles would be much harder to eradicate.

9.4 Pathogens and Networks

The SIS model approach is to course a first approximation of epidemics, although it allows us to understand many interesing cases. But the population biology of infectious diseases involves further complexities (Anderson and May 1991). These include spatial dynamics (Grenfell et al. 2001), evolutionary change (Earn et al. 2002), and the structure of human-based networks (Pastor-Satorras and Vespignani 2000; May and Lloyd 2001; Colizza et al. 2007). As mentioned at the end of chapter 5, the organization of real networks is usually heterogeneous. This includes several related webs, particularly two: transportation networks and social interaction networks. The first group is dominated by airports: a given individual can travel from one part of the globe to another very easily. Most airports have one or two links, and a few dominate the scene. These hubs connect to many parts of the world and they largely influence the likelihood of a given infectious disease to propagate worldwide as a pandemic. Moreover, on the social level, sexual contact networks have been shown to display scale-free organization too (Liljeros et al. 2001).

The implications of this pattern have been widely discussed because of their potential use in controling the spread of both biological and computer viruses (Lloyd and May 2001). In a nutshell, both large airports and promiscuous sexual partners need to be taken into account. The first is crucial because it allows pathogens to jump to multiple, geographically distant targets. The second is significant because it represents the highest risk group, given its leading role in allowing pathogens to be transmitted. The presence of this risk group modifies one of our main findings: it reduces or even removes the presence of an eradication threshold (Pastor-Satorras and Vespignani 2001). This lack of a phase transition implies that epidemics will always propagate due to the many opportunities provided by highly connected nodes. We can return again to our original scenario with a threshold by properly influencing the hubs (Dezso and Barabási 2002; Pastor-Satorras and Vespignani 2002; Zanette and Kuperman 2002; Latora et al. 2006; Colizza et al. 2007).

10

GENE NETWORKS

10.1 Genes and Cell Types

Any multicellular organism, from sponges to humans, is formed by a vast amount of cells defining an integrated being (Wolpert 2004). Cells are the fundamental units of life, and within a complex organism, they cooperate in different ways, forming tissues and organs and sustaining the life of the individual in a reliable and homeostatic manner. From a single initial cell, a multicellular organism develops into a complex life form with different cell types (figure 10.1a). Cell types emerge through the process of *differentiation* (Alberts et al. 2004). Such process involves the interactions between a cell's gene regulatory networks and the information sent by neighboring cells and the environment.

How are different cell types generated? Mounting evidence points toward genetic switches of different types as the underlying mechanism of cell differentiation (Huang et al. 2005). Switching mechanisms would be responsible for creating distinct patterns of gene activation and repression. If cells are in physical contact or can communicate through diffusion, nonlinear mechanisms of cell-cell information exchange can propagate signals through space and lead to a stable spatial pattern. The regulatory pattern of interactions is expected to be complex and involve both positive

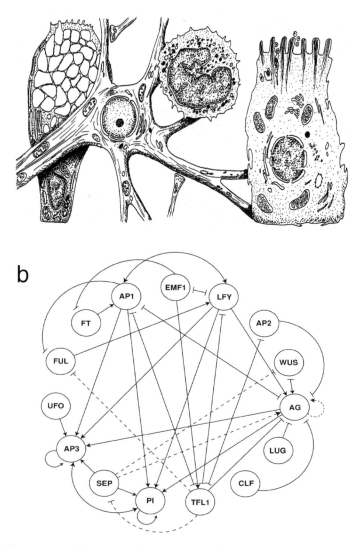

a

b

Figure 10.1. Tissues and organs are formed by combinations of different cell types (a) displaying very diverse morphologies. Each type of cell displays a number of specific functionalities

(continued)

and negative links. An example of a given subnetwork is shown in figure 10.1b (Espinosa-Soto et al. 2004). This network involves regulatory interactions associated with the determination of cell fate in the flowering of a species of plant, *Arabidopsis thaliana*. Untangling the wires of such complex networks is one of the main challenges in systems biology (Bornholdt 2002, 2005; Alvarez-Buylla et al. 2007).

This problem involves many degrees of freedom. Moreover, it is remarkable that the possible and the actual in the universe of *potential* cell types are very different. Each cell type can be understood (on a first approximation) as a set of ON-OFF genes: some genes will be activated, generating special proteins characteristic of the given cell type, whereas others will be inactivated (figure 10.1b). Each cell shares the same set of N genes (from thousands to tens of thousands) and thus the *potential* number of cell types N_c under this ON-OFF picture will be astronomic, namely:

$$N_c \sim 2^N \qquad (10.1)$$

whereas in reality it remains confined to a much smaller number, from tens to hundreds (Kauffman 1993; Carroll 2001). Why is this number so small? More important, cell types are stable states, not easily able to switch from one to another without

---

Figure 10.1 (continued). and cooperation among cells allow tissues and organs to perform optimaly. Each one seen, as a first approximation, as the result of the activation of a given set of genes (drawing by R. Solé). In (b) we show an example of a gene network involved in the determination of cell states of the floral organ of the plant *Arabidopsis thaliana*. Here network nodes represent active proteins of corresponding genes, and the edges represent the regulatory interactions between node pairs (arrows are positive, and blunt-end lines are negative). Dashed lines are hypothetical interactions. (Redrawn from Espinosa-Soto et al. 2004)

some perturbation. On the contrary, cellular states tend to be stable over time, consistent with the desired stability for a complex organism to survive.

10.2 Boolean Dynamics

How to deal with the dynamics of cells at the genetic level, if the number of genes and attractors is so high? One possible shortcut is given by Random Boolean Networks (RBN), a class-of-discrete dynamical system where each element is synchronously updated through a specified function:

$$S_i(t + 1) = \Phi_i[S_{i_1}(t), S_{i_2}(t), \ldots, S_{i_K}(t)] \qquad (10.2)$$

where $i = 1, \ldots, N$ and $S_i(t) \in \Sigma \equiv \{0, 1\}$ are the only two possible states (ON-OFF). Let us note that each element has a *different* associated function, since different sets of genes send input to it (that's why we use an index for each Φ function). The previous definition assumes that time is discrete (states are updated using some clock having a characteristic time scale) and that each gene receives input from (is regulated by) exactly K other genes (where $K = 0, 1, \ldots$). Many potential choices can be made in defining the exact Boolean functions Φ_i. The easiest approach is to choose each randomly from the set of all K-inputs Boolean functions. In other words, for each K-string of possible inputs, we randomly choose the output (0 or 1) with some probability p. The links between elements are also chosen at random (as in a random graph, chapter 5).

Gene networks able to generate multiple attractors have been identified in real systems (Buylla et al. 2007). An example is given in figure 10.1b, where we display the gene-gene interaction map associated to early flower development (Chaos et al. 2006). In this case, six cell types are matched by six attractors obtained from a Boolean representation of gene states. These models exhibit a range of dynamical patterns, from fixed points to oscillations and chaos (Aldana et al. 2003; Drossel 2007). Fixed points are sets $S^* = (S_1^*, \ldots, S_N^*)$ such that

$$S_i^* = \Phi_i[S_{i_1}^*, S_{i_2}^*, \ldots, S_{i_K}^*] \qquad (10.3)$$

for all $i = 1, \ldots, N$. A simple example of these networks is shown in figure 10.2a–b, where a small $N = 3, K = 2$ case is being considered. Here the functions used are AND and OR. These Boolean functions are simply defined by means of Boolean tables, namely:

$S_2(t)$	$S_3(t)$	$S_1(t+1)$
0	0	0
0	1	0
1	0	0
1	1	1

$S_1(t)$	$S_3(t)$	$S_2(t+1)$
0	0	0
0	1	1
1	0	1
1	1	1

for the AND and the OR gates associated to S_1 and S_2, respectively.

Using these tables, it is easy to compute the next state exhibited by the whole network (Kauffman 1993). Even in this simple example, we can see that the possible dynamical patterns are nontrivial. In figure 10.2c we show the possible transitions between different states. If the system starts at $(0, 0, 0)$, it remains there (point attractor) whereas if it starts from $(0, 0, 1)$ or $(0, 1, 0)$, these two states flip into each other defining a cycle. The rest of the states flow toward the second fixed point $(1, 1, 1)$ thus defining a basin of attraction.

10.3 Percolation Analysis

The previous example is just a picture illustrating how the dynamics is defined. When N becomes large, the combinatorial possibilities rapidly explode and it does not make much sense to use tables. However, from a statistical point of view, two

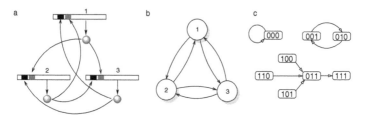

Figure 10.2. Boolean networks and attractors. In (a) a hypothetical gene network is shown, where each gene (long rectangles) produces a protein (gray balls) that regulates the other two genes. An abstract formulation of this small genetic circuit is shown in (b) where the explicit mechanism of gene-gene interaction is replaced by arrows indicating the existence of an interaction (here $N = 3$ and $K = 2$ is used). For a given choice of the Boolean functions (an OR gate for genes 2 and 3 and an AND gate for 1, see text), the possible transitions between states for the state space $\{S_1, S_2, S_3\}$ are shown in (c). Three attractors are obtained.

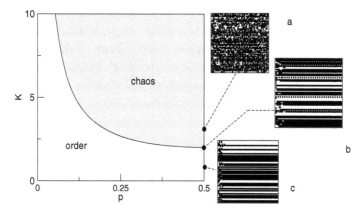

Figure 10.3. Phase transition in random Boolean networks. The top left plot displays the two phases in the connectivity-bias parameter space. Three examples of the spatiotemporal behavior are indicated in (a–c) for three different values of K at $p = 0.5$. Here time goes from left to right, and black and white squares indicate 1 and 0 states, respectively.

well-defined phases have been identified (Derrida and Stauffer 1986; Derrida and Pomeau 1986; Kurten 1988; Kauffman 1993; Glass and Hill 1998; Aldana et al. 2003; Drossel 2005). Assuming that the probability of having a zero or one in the Boolean table is the same (i.e., we generate the two states using a fair coin), we can tune the number of connections K. For $K = 1$, the system is frozen: every element rapidly achieves a given steady state and remains there. For $K \geq 3$ the system is chaotic, with elements changing in a seemingly random fashion. This can be further generalized by introducing an additional parameter p that measures the probability that the output of equation (2) is zero.

These two phases describe two different scenarios involving order and disorder. Moreover, the attractors in each phase are highly sensitive to perturbations: in both cases, a change in the state of a given element generates changes in a system's state. At the ordered domain, an element that is randomly chosen and changed remains in the same state (the input from others is too weak and unable to reverse the flip). Instead, in the disordered phase, a flip triggers a chain reaction that propagates through the entire system (too many connections). In both cases, a very large number of attractors is present. However, for $K = 2$ the system displays a reduced number of attractors that is highly stable: a given perturbation is typically reversed and thus homeostasis is at work. Such properties are compatible with what we would expect for a cell type, as discussed earlier.

The dynamical equations describing the RBN model (2) can be seen as a set of tables. For each input, we have a given output:

$$S_{i_1}(t), S_{i_2}(t), \ldots, S_{i_K}(t) \longrightarrow S_i(t+1) \qquad (10.4)$$

and each value $S_i(t+1)$ (i.e., the right column in the table) is either zero or one with some probability p. In other words, we have, for any S_i in the system,

$$P[\Phi_i(S_{i_1}(t), S_{i_2}(t), \ldots, S_{i_K}(t)) = 1] = p \qquad (10.5)$$

and

$$P[\Phi_i(S_{i_1}(t), S_{i_2}(t), \ldots, S_{i_K}(t)) = 0] = 1 - p \qquad (10.6)$$

This parameter is the so-called bias and it seems clear that it will also influence the location of the phase transition.[1] In our previous example, we used $p = 1/2$, but a very small (or large) p will clearly make the system much more homogeneous and thus more likely to be ordered, no matter the value of K.

In order to derive the mean field phase transition line, we need only consider a simple percolation argument (Luque and Solé 1997). At the subcritical phase, a change in a given element is unable to propagate through the system, whereas within the chaotic phase any change has an impact. This reminds us of the percolation phenomenon discussed in chapter 4: provided that a critical threshold is reached, a single flip in the network can generate a cascade of changes. In our context, a flip means that we change one input element, namely:

$$\{S_{i_1}(t), \ldots S_{i_j}(t) \ldots, S_{i_K}(t)\} \longrightarrow \{S_{i_1}(t), \ldots 1 - S_{i_j}(t) \ldots, S_{i_K}(t)\}$$
$$(10.7)$$

What is the probability that such change generates further changes in the other elements that receive input from S_i? In figure 10.4a the basic situation is shown using a simple tree structure. Here at the bottom we have the element that is in a given state (1 in our example). This element send inputs to others (an average of $\langle K \rangle$) which are also in a given state. These send further inputs to others. If we flip this element (i.e., $1 \rightarrow 0$, figure 10.4b), what is the likelihood that this will trigger an avalanche?

If we choose an arbitrary element, the question is what is the probability that the output in the table defined by Φ_i will change

---

[1] For the small examples defined above and used to generate the attractors of figure 10.2, we have $p(AND) = 1/4$ and $p(OR) = 3/4$. But no extrapolations from these small-sized examples can be made in terms of the global behavior of large systems.

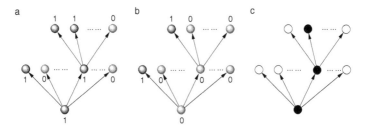

Figure 10.4. Propagation of perturbations in random Boolean networks. In (a) we show part of a given network using a treelike representation, downstream from a given element. This element (bottom) send inputs to K other elements which send inputs to others. If we flip the state of the bottom element (b) further changes can be induced, provided that the Boolean tables allow this to happen. In (c) we plot in black the propagating changes, which can percolate if the appropriate critical condition is reached (see text).

to a different value. This is easily found: if the initial output is 1, the probability that it changes to zero is simply $p(1 - p)$. Since we need to consider the symmetric situation $(0 \rightarrow 1)$, we actually have a total probability of change $2p(1 - p)$. Every element affects $\langle K \rangle$ others, and thus the total number of changes reads

$$\mathcal{N} = 2p(1 - p)\langle K \rangle \qquad (10.8)$$

Now we are ready to find the critical line: the boundary separating propagation from a frozen state is simply $\mathcal{N} = 1$, which gives:

$$K_c(p) = \frac{1}{2p(1 - p)} \qquad (10.9)$$

and for the special case $p = 1/2$, we have $K_c(1/2) = 2$.

The previous formalism can be easily applied to more complex situations, including random networks with multiple states and neural networks (Luque and Solé 1997).

10.4 Is Cell Dynamics Boolean?

The assumption that gene network dynamics can be represented in terms of Boolean, discrete sets of states, is of course a strong one. Such an approach is necessarily a crude approximation of real genes. But we also need to keep in mind that the level of approximation used is directly related to the questions we are trying to answer.

The correctness of the ON-OFF picture has received considerable attention over the years (see Bornholdt 2008 and references cited; Macia et al. 2009). The truth is that the logic of gene regulation has much in common with a computational device (Bray 1995; Hogeweg 2002; Macia and Solé 2008), and switchlike genetic behavior is known to be more the rule than the exception. In order to illustrate this point, let us consider the simplest example of a genetic switch, using a single gene coding for a protein P that activates the gene through a positive feedback loop (figure 10.5a). Such a scenario has been shown to occur experimentally by tuning the strength of transcription of a given gene (Isaacs et al. 2003). The protein is assumed to form a dimer in order to perform its regulatory function. If p indicates the concentration of P, it can be shown that the kinetic description of this system is given by the nonlinear equation:

$$\frac{dp}{dt} = \frac{\alpha p^2}{1 + p^2} - \delta p \qquad (10.10)$$

where δ is the degradation rate of monomers.

Together with a trivial fixed point $p^* = 0$, which is always stable, two additional points are obtained, namely

$$p^*_\pm = \frac{1}{2}\left[\frac{\alpha}{\delta} \pm \sqrt{\left(\frac{\alpha}{\delta}\right)^2 - 4} \right] \qquad (10.11)$$

The higher is stable whereas the lowest is unstable. If we use α/δ as a bifurcation parameter, we can see that these two fixed

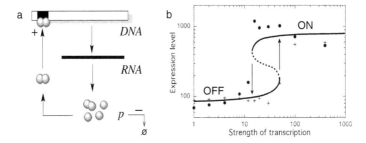

Figure 10.5. In (a) a simple example of a genetic circuit displaying two possible alternative states is shown. Here a gene is expressed and translated into a protein *P* (gray balls). This protein degrades and is removed (empty ball) or forms dimers which can bind to the DNA and trigger further synthesis. Experimental results using engineered systems (b) confirm that this system is able to behave as a switch (adapted from Isaacs et al. 2003).

points will exist provided that $\alpha > 2\delta$. In other words, two basic alternative states exist if the previous inequality holds. The one chosen might depend on how other cellular signals tune α or δ.

The previous example illustrates our point: although strictly speaking the underlying system is continuous, the nature of the final decisions can be understood in terms of switching logic. Moreover, we might ask why bistable, Boolean dynamics is so widespread: why not tri-, four-, or n-stable? Although tristable dynamics might be involved in some special circuits, such as stem-cell switches (Huang et al. 2007), it seems that, beyond two alternative states, switches become less likely to properly work in a reliable way (Macia et al. 2009). We could say that nature has no choice but being (nearly) Boolean.

11

CANCER DYNAMICS

11.1 Abnormal Growth

Cancer is the result of a system breakdown that arises in a cell society when a single cell (due to a mutation or set of mutations) starts to display uncontrolled growth (Hanahan and Weinberg 2000; Weinberg 2007). The cooperation that maintains the integrity of a multicellular organism is thus disrupted. Further changes in the population generated by such abnormal cell growth can lead to malignant tumor growth, eventually killing the host. From an evolutionary point of view, tumor progression is a microevolution process in which tumors must overcome selection barriers imposed by the organism (Merlo et al. 2006). The emergence and evolution of tumors involve a number of phenomena that are well known in complex systems (Kitano 2004; Anderson and Quaranta 2008). These include, for example, growth, competition (Michelson 1987; Gatenby 1996), spatial dynamics (González-Garcia et al. 2002; Cristini et al. 2005), and phase transitions (Garay and Lefever 1977; Solé 2003; Solé and Deisboeck 2004).

As discussed in (Alberts et al. 2007), a multicellular system is not far from an ecosystem whose individual members are cells, reproduced in a collaborative way and organized into tissues. In this sense, understanding multicellularity requires concepts that are well known in population dynamics such as birth, death, habitat, and the maintenance of population size (Merlo et al. 2006). Under normal conditions, there is no need to worry about selection and mutation: as opposed to the survival of the fittest, the cell society involves cooperation and, when needed, the death of its faulty individual units. Mutations occur all the time but sophisticated mechanisms are employed in detecting them and either repairing the damage or triggering the death of the cell displaying mutations (Weinberg 2007). Abnormal cells can be indentified from within (i.e., through molecular signaling mechanisms operating inside the damaged cell) or by means of interactions with other cells. The latter mechanism involves immune responses.

Selection barriers (such as an attack from the immune system or physical barriers of different types) can be overcome by a tumor provided that the diversity of mutant cells is high enough to generate a successful strain. High mutation rates are thus a way to escape from the host responses and it is actually known that most human cancers are *genetically unstable* (Weinberg 2007). Genetic instability results from mutations in genes that are implicated in DNA repair or in maintaining the integrity of chromosomes. As a result, mutations accumulate at very high rates.

Modeling cancer is far from trivial (Wodarz and Komarova 2005; Spencer et al. 2006; Byrne et al. 2006; Anderson and Quaranta 2008). Although the first impression is that tumors are simply growing, disordered pieces of tissue, the reality is that they involve several levels of complexity. Cancer is a complex adaptive system and as such exhibits robustness (Kitano 2004; Deisboeck and Couzin 2009) and thus adaptability to changing conditions. However, as any other complex system, cancer can also display Achilles' heels. In this context, models offer the possibility of exploring potential sources of fragility such as catastrophes and breakpoints. Two examples are presented below.

11.2 Tumor Decay under Immune Attack

As mentioned above, cancer cells must overcome different types of selection barriers in order for tumors to expand. Among these barriers, the immune system (IS) appears as a specially important one. The immune system is itself a complex system, able to respond to the invasion of foreign pathogens with high efficiency (Janeway et al. 2001). There is no space to provide even a basic overview of how the IS works. We will confine ourselves to a rather important component and how it interacts with cancer. In a nutshell, the IS response is based on the recognition of a very large class of molecules, also known as *antigens*. These are mostly carried by infectious agents, such as viruses and bacteria.[1] Immune responses are directed toward detecting and eliminating these molecules and their carriers. However, if the carrier is able to change itself (such as RNA viruses, see chapter 6), it can escape from the immune attack. Different escape strategies are known to occur, all requiring some degree of internal plasticity.

Among the key mechanisms at play, an important part of the IS response is based on special types of cells able to detect anomalous molecular markers at the surface of infected cells. These include the cytotoxic T cells (CTLs). Each class of CTL is able to recognize one specific antigen among all the huge set of possible molecules. Once recognition occurs, an attack is launched and the CTL injects toxic granules into the target cell (figure 11.1) eventually killing it.

In principle, these mechanisms could be a source of cancer-cell removal. However, tumor cells can evade IS recognition in different ways (Weinberg 2007), and sometimes the IS becomes

---

[1] However, the IS needs also to distinguish between nonself and self: it must avoid attacking normal cells. The IS actually learns to distinguish what is part of the body, but such recognition is not perfect and mistakes can ocur. Such mistakes happen from time to time causing different types of so-called autoimmune diseases.

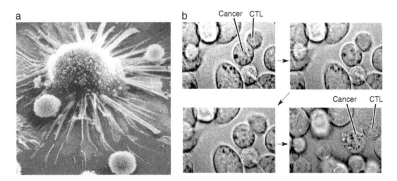

Figure 11.1. Attack on a cancer cell unleashed by a CTL lympho-cyte, as shown in (a) where the larger cell belongs to a neoplasic tissue and the small spheres are cytotoxic cells. In (b) the sequence shows the whole process, including the formation of a cell-cell complex and the final killing of the tumor cell (adapted from Weinberg 2007).

ineffective against cancer growth. The interaction between cancer and the immune system actually involves multiple layers of complexity (Adam and Bellomo 1997). An especially important problem is determining the conditions under which the IS will be able to remove cancer, and some models reveal the presence of transition phenomena. Below we present the Garay-Lefever model of cancer–immune system interactions (1977), which is shown to display both first- and second-order transition phenomena (Garay and Lefever 1977).

Consider a population of tumor cells, to be indicated as X, that can be attacked by some class of cytotoxic killer cells (T-cells). Here cancer cells replicate at a given rate η. In this model, T cells can recognize and kill cancer cells by detecting special markers on the tumor cell surface. The identification occurs at a rate k_1 and triggers the formation of a cell-cell complex C (figure 11.1). Once the cancer cell has been killed, the complex dissociates and the T cells are free again.

The previous components define a set of reactions, namely:

$$X \xrightarrow{\eta} 2X \qquad (11.1)$$

for cancer growth kinetics and

$$T + X \xrightarrow{k_1} C \xrightarrow{k_2} T + D \qquad (11.2)$$

for cancer-CTL interaction. Here D indicates a dead (cancer) cell. The rates k_1 and k_2 measure the speed at which the complex forms and disintegrates, respectively. The corresponding equations describing this dynamics are:

$$\frac{dX}{dt} = \eta X \left(1 - \frac{X}{K} \right) - k_1 TX \qquad (11.3)$$

$$\frac{dT}{dt} = k_2 C - k_1 TX \qquad (11.4)$$

where a maximum number of cancer cells K is assumed to exist. The first term on the right-hand side of the cancer-growth equation is thus consistent with a logistic growth model (chapter 2).[2]

As defined here, cytotoxic cells act as predators, reducing the number of cancer cells at a rate $k_1 TX$, that is, depending on the encounter rates, efficiency of recognition, and other phenomena captured by k_1. Moreover, killer cells forming complexes get back into the free T cell pool as soon as the cancer cell is killed at a rate k_2.

This two-dimensional system can be reduced by further assuming (Garay and Lefever 1977) that the lifetime of the C complex is short and thus a steady state assumption:

$$\frac{dT}{dt} \approx 0 \qquad (11.5)$$

can be made. This equilibrium condition just tells us that cytotoxic cells form complexes that rapidly split (compared to

---

[2] Actually, a more realistic representation of tumor growth dynamics is provided by the so-called Gompertz model, where the kinetic equation is given by $dx/dt = \eta X \log(K/X)$.

other phenomena involved in the model). The total number of cytotoxic cells, E, can be computed as follows

$$E = T + C \qquad (11.6)$$

since E includes both complex-associated and free killer cells. In the following, we assume that E is a conserved (constant) value.

Using the previous stationary approach for T cell dynamics, we have $T(X) = k_2 C / k_1 X$ and using $C = E - T$, we obtain:

$$T(X) = \frac{E}{1 + \frac{k_1}{k_2}X} \qquad (11.7)$$

which allows us to write down the whole mean field equation for tumor growth dynamics in terms of a one-dimensional system,[3] namely:

$$\frac{dX}{dt} = \eta X \left(1 - \frac{X}{K}\right) - \frac{k_1 X E}{1 + \frac{k_1}{k_2}X} \qquad (11.8)$$

Following Garay and Lefever, we will reduce the number of parameters in this model by using the following rescaled quantities: $\mu = k_1 E / \eta$, $\theta = k_2 / k_1 N$ and $x = k_1 X / k_2$ (time t would be rescaled accordingly as $t \rightarrow \eta t$, but we keep the same notation). The model now reads:

$$\frac{dx}{dt} = x(1 - \theta x) - \frac{\mu x}{1 + x} \qquad (11.9)$$

Using this reduced one-dimensional model, three possible fixed points are obtained (figure 11.2), namely $x^* = 0$ (cancer extinction) and the two nontrivial solutions:

$$x_{\pm} = \frac{1}{2\theta} \left[1 - \theta \pm \sqrt{(1 + \theta)^2 - 4\theta\mu}\right] \qquad (11.10)$$

The tumor-free fixed point will be stable provided that

$$\lambda_\mu(0) = \left(\frac{\partial \dot{x}}{\partial x}\right)_{x=0} < 0 \qquad (11.11)$$

---

[3] The resulting model is very similar to the one obtained by considering the interaction between cancer cells and a given anticancer drug.

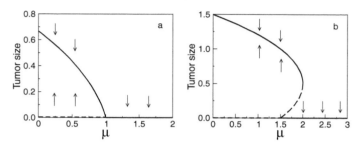

Figure 11.2. Bifurcations in the Garay-Lefever model. Here we plot the stable solutions (thick, continuous lines) for (a) $\theta = 1.5$ and (b) $\theta = 0.5$ corresponding to the monostable and bistable domains (see text).

which in our case is simply $\mu > \mu_c = 1$. Since the parameter μ captures the conflict between the rates η (tumor cell formation) and k_1 (tumor cell removal) the critical condition $\mu_c = 1$ naturally separates cancer survival from extinction.

Looking at the nontrivial solutions, we can see that two possible scenarios can be defined:

1. For $\theta > 1$, the fixed point x_- will be negative (and thus biologically irrelevant) whereas x_+ remains positive provided that $\mu < 1$. In this case, a cancer cell population will exist below the critical point μ_c, exchanging stability with the trivial solution.

2. For $\theta < 1$, when the inequality $(1 + \theta)^2 \geq 4\theta(\mu - 1)$ holds, the two nontrivial solutions exist. It is not difficult to see that, given the previous results, the interval of existence of these two simultaneous solutions is:

$$1 < \mu < \frac{(1 + \theta)^2}{4\theta} = \bar{\mu} \qquad (11.12)$$

defining a domain of bistable behavior. For $\mu < 1$ only the first solution x_+ is present. For the bistable domain, we can describe the two solutions as follows:

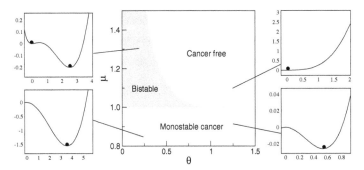

Figure 11.3. The three phases exhibited by the Garay-Lefever model (1977). Here the upper white domain indicates that no cancer cells are stable (tumor extinction) whereas two other areas correspond to the presence of monostable cancer (white low domain) and two alternative states (gray area). Four examples of the associated potentials are also shown, with filled circles indicating the presence of stable states.

$$x_\pm = \frac{1}{2\theta} \left[1 - \theta \pm \sqrt{\theta(\bar{\mu} - \mu)} \right] \qquad (11.13)$$

In figure 11.2 we show two examples of the bifurcation diagrams associated with the model. In (a) an example of the continuous transition scenario is obtained for $\theta > 1$ whereas the bistable case, for $\theta < 1$, is shown in (b). In figure 11.3 the basic results are summarized by representing the three phases on the (θ, μ) parameter space. We also plot four examples of the associated potential function, here given by

$$\Phi_\mu(x) = -\frac{x^2}{2} + \frac{\theta x^3}{3} + \mu(x - \ln(1 + x)) \qquad (11.14)$$

which also illustrate the exchange of stability between different attractors.

One especially interesting result of this model is the existence of a catastrophic shift from cancer to cancer-free tissue. Looking at the transition shown in figure 11.2b, we conclude (under our special assumptions) that a slight increase in μ could trigger the remission of cancer and eventually its disappearance. Such phenomena have actually been reported in some rare cases where advanced (usually lethal) tumors were shown to experience spontaneous reversal (Cole 1981). These spontaneous remissions have also been observed in some animal models (Cui et al. 2003). The mechanisms pervading the dynamics of these remarkable events are largely unknown, but a late, powerful immune response is likely to be involved (see Cui et al. 2003; Weinberg 2007). The almost sudden regression experienced by these subjects seems consistent with the first-order shifts displayed by our toy model.

11.3 Thresholds in Cancer Instability

In this section we consider a different scenario in which thresholds to cancer viability might exist. The key property now is genetic instability. More precisely, we consider the growth of tumor populations in those types of cancer where mechanisms involved in the maintenance of genomic integrity have failed (Lengauer et al. 1998; Hoeijmaker 2001; Weinberg 2007). Such failure is linked to mutations in a class of genes responsible for detecting and controlling potential mistakes in the process of DNA replication. This leads to the so-called mutator phenotype (Loeb et al. 2003). If DNA stability is jeopardized, increasing levels of instability foster tumor progression and further instability (see Cahill et al. 1999; Loeb et al. 2003). This leads to the question of what levels of instability can be achieved before harmful mutations stop further progression. Such instabilities are observable at different levels, from chromosomal aberrations to pathological tissue organization. The latter is well illustrated in figure 11.4, where different stages of colon cancer are shown at different times of tumor development (Gatenby and Frieden 2002). A clear loss of order can be appreciated, although some

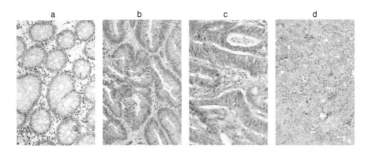

Figure 11.4. Degradation of tissue organization at different stages of the evolution of a colon cancer (adapted from Gatenby and Frieden 2002). Here we have: (a) normal tissue (b) colon adenomatous polyps with low grade, (c) dysplasia, and (d) undifferentiated, invasive colon cancer with highly disordered morphology and loss of normal colon tissue organization.

degree of coherence is observable as cells maintain some capacity to differentiate. Eventually (as illustrated by the last picture in the sequence) a complete loss of structure occurs. It is known that many of these cancer cells are actually not able to replicate, thus suggesting that (as it occurs with RNA viruses, see chapter 7) some thresholds of viability might be present (Cahill et al. 1999; Solé 2003; Solé and Deisboeck 2004). If true, then such thresholds could be exploited for cancer treatment. Below we present a minimal model of cancer growth illustrating these ideas.

Assuming that two populations of cells are at play, namely normal (n) and cancerous (x), we will introduce a simple, two-dimensional model of cancer instability. The model assumes that both populations are homogeneous, thus all cells in each subset are identical in their replication capacities. This is of course an oversimplification, particularly when dealing with unstable cancer, where by definition high diversity is expected. Assuming that these populations compete for available space and resources, the simplest model reads:

$$\frac{dn}{dt} = rn - nF(n, x) \qquad (11.15)$$

$$\frac{dx}{dt} = r\Gamma(\mu)x - xF(n, x) \qquad (11.16)$$

where r is the replication rate of normal cells in the original tissue and μ indicates the rate of genetic instability.[4] The function $\Gamma(\mu)$ includes the (nonlinear) effect of instability on the growth of unstable cancer cells. Although μ can be understood as a mutation rate, other changes beyond point mutations are being taken into account (such as chromosomal rearrangements). The last terms on the right-hand side of both equations introduce an outflow that is required to guarantee a finite population size.

For simplicity, we assume that there is a maximum population size and that the total population is constant, that is, $n + x = C$, which for simplicity we fix to $C = 1$. This is the constant population constraint used in chapter 6, which implies that $dx/dt + dn/dt = 0$. Using this constraint, we obtain:

$$F(n, x) = rn + r\Gamma(\mu)x \qquad (11.17)$$

And since $n = 1 - x$, we can reduce our previous model to the following one-dimensional system:

$$\frac{dx}{dt} = r(\Gamma(\mu) - 1)x(1 - x) \qquad (11.18)$$

which is nothing but a logistic equation (chapter 2). We know from our previous analysis that two fixed points are present: the zero-population one $x^* = 0$ and the maximum population state, here $x^* = 1$. It is easy to see that the first is stable if $\Gamma(\mu) < 1$ and unstable otherwise. By properly defining the function $\Gamma(\mu)$ we might be able to define the conditions under which genetic

[4] Previous models used a formal approach based on the quasispecies model developed in chapter for RNA viruses.

instability allows cancer growth to occur and overcompete the host tissue.

In order to propose a reasonable form for $\Gamma(\mu)$, two features of instability must be considered. The first is that, at low levels, instability allows the emergence of clones able to grow faster and overcome selection barriers. In other words, some amount of instability is good for the cancer population: it provides a source for adaptability and selective advantages. The second deals with the negative part of the story: at high levels, mutations tend to be harmful and cells nonviable. The function should be such that $\Gamma(0) = 1$ (i.e., no effect is at work) whereas we should expect $\Gamma(\mu) \to 0$ as instability grows (i.e., $\mu \to \infty$). One possible choice (Solé et al. 2008) is using:

$$\Gamma(\mu) = (1 + \alpha\mu)e^{-\mu/\mu^*} \qquad (11.19)$$

which is consistent with all the previous requirements. Here $\alpha \geq 0$ is a parameter that weights the selective advantage of genetic instability: the higher its value, the faster the increase in growth rate. The parameter μ^* is simply some characteristic rate that we fix here to $\mu^* = 0.5$. By using this functional form for $\Gamma(\mu)$ we find two phases, as indicated in figure 11.5. Since the two alternative possibilities ($x^* = 0$ and $x^* = 1$) are separated by a critical line, a sharp change is expected to take place near this boundary. The corresponding potential is now:

$$\Phi_\mu(x) = -r(\Gamma(\mu) - 1)\left(\frac{x^2}{2} - \frac{x^3}{3}\right) \qquad (11.20)$$

The prediction from this model, and related ones (see Solé 2003; Solé and Deisboeck 2004; Solé et al. 2008), is that unstable tumors close to the critical boundary could be highly adaptive but also fragile with respect to small variations of instability levels. Although our model is based on fixed parameter values, more detailed models involving changing instability levels reveal that such levels can increase under Darwinian dynamics,

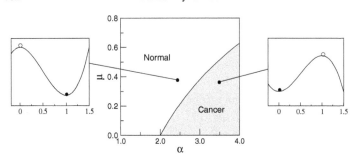

Figure 11.5. Phases in unstable cancer dynamics. The gray area indicates the domain where unstable cancer overcompetes the host tissue. When close to the separation boundary, small changes of instability levels can generate a shift in the behavior (provided that normal cells are still present). Two examples of the associated potentials are shown, with $\mu = 0.4$ and $\alpha = 2.5$ (left) and $\alpha = 3.5$ (right), respectively.

pushing the cancer population close to the critical domain. If further instability were introduced (as might be the case while using radiation or appropriately designed drugs) the tumor could eventually cross the line and decay.

11.4 Cancer as an Evolving System

As pointed out by Merlo et al. (2006), cancer is a disease of clonal evolution within the body. Although such an idea was formulated early (Cairns 1975), modeling and clinical and molecular data reveal that evolution and genomic instability in particular, are key to carcinogenesis. Theoretical approaches to the origins and development of cancer need to take such considerations into account.

The previous models have several important limitations. One is their lack of explicit genetic variability and heterogeneity. We have already mentioned that tumors are genetically unstable in terms of the diversity of cellular populations. This has been shown in many different studies, tracking tumors both in space and time (Gonzalez-Garcia et al. 2002; Maley et al. 2006). Since spatial patterning is a source of opportunities for diversification (Solé and Bascompte 2007), appropriate models should include spatial degrees of freedom (by reducing the impact of competitive interactions and favoring cooperation) in their ecological description of tumors. Moreover, there is also an evolutionary element to consider: tumor progression takes place to a large extent by exploiting the evolutionary legacy affecting our genomic organization. Appropriate models need to consider this feature (Spencer et al. 2006). Finally, although our picture of tumor dynamics is mainly sustained by the consideration of competitive interactions, cooperation might also play an important role in driving cancer progression (Axelrod et al. 2007).

12

ECOLOGICAL SHIFTS

12.1 Change in Ecology

Ecosystems are complex and adaptive (Milne 1998; Levin 1998; Solé and Levin 2002; Pascual 2005), displaying nontrivial forms of organization in both space and time. An important component of change in ecological systems involves the rapid response of some communities to slow changes in external variables (May 1977; Lefever and Lejaune 1997; Scheffer et al. 2001; Scheffer and Carpenter 2003; Rietkert et al. 2004). It has been shown in different contexts that constant changes in water availability, decline of some plant or fish species or increasing temperatures can trigger sudden changes in ecosystem organization. The outcome of such changes is often a shift from one stable state to another. This can affect a large geographic area, as happened with the transition from green to desert in the Sahara, and discussed in chapter 3. Such changes are the result of nonlinearities coupling external inputs and internal responses. In terms of dynamical systems, populations can shift suddenly from one state to another as a limiting input parameter changes beyond some given threshold.

A perfect illustration of the type of sudden changes that can occur in complex ecosystems is provided by the dynamics of semiarid vegetation. Semiarid and arid habitats represent

30 percent of the current ecosystems on our planet, and are strongly limited by water availability. Desertification is an increasing threat to many habitats in the world, having a great impact on biodiversity and productivity. Such a process is being accelerated by raising human intervention through fires, grazing, habitat loss and fragmentation, and decreased water availability. Both studies of ancient paleoecosystems (see Foley et al. 2003 and references cited) and current arid ecosystems (Kefi et al. 2007; Scanlon et al. 2007; Solé 2007) reveal the presence of complex dynamics coupling water and vegetation.

Many other examples of qualitative changes have been reported in the literature. Among them, shifts between alternate states in shallow lakes provide another compelling example (Scheffer et al. 1993). In these systems, the unperturbed, pristine state with clear water and submerged vegetation becomes replaced by a new state characterized by turbid water (mainly due to phytoplankton accumulation). This occurs as a consequence of a human-induced increase in eutrophication associated with excess nutrient concentration. It was early observed that water remains crystal clear under increased nutrient uptake until a threshold is passed. Afterward, a shift from clear to turbid occurs in an abrupt fashion. Reduction of nutrient availability seldom reverses this state, thus suggesting that memory effects (see chapter 3) might be at work.

Another example of ecosystem change leading to extinction under the presence of some thresholds is provided by habitat loss and fragmentation. Current rates of habitat destruction are extremely high. Habitat loss strongly enhances the detrimental effects triggered by species introductions, pollution, climate change, and hunting. The physical changes associated with habitat loss and fragmentation include reduction of total area and productivity of native areas, isolation of forest remnants, and changes in the physical conditions of the remnant fragments. These antropogenic changes trigger further community responses that sometimes end in a biotic collapse (Wilcove 1987). Multiple imbalances arise, leading to nonlinearities and cascade effects throughout the ecological webs. The loss of key species can promote subsequent loss of their predators, parasites, and mutualists (Terborgh et al. 2001). Given the magnitude and consequences of habitat destruction, it is imperative to get enough insight to understand the effects of habitat loss on species survival, and predict its further consequences. Since economic trade-offs are at play, scientists are often faced with the question of how much habitat can be destroyed before a certain species goes extinct. In order to fully understand the consequences of habitat loss, models play a relevant role, particularly in forecasting the effects of landscape degradation on web structure and stability.

Sudden transitions in ecosystems are likely to be common. Such tipping points represent an important problem, not only in the context of ecology but also in other fields, including medicine, geology, and finance (Scheffer 2009). One obvious question is, how predictable are they: can we detect warning signals? (Scheffer et al. 2009). Below we consider two examples of critical transitions in arid ecosystems and outline the problem of early signal detection.

12.2 Green-Desert Transitions

We choose here a very simple type of model that has been introduced in two different forms in the literature (Klausmeier 1999; Shnerb et al. 2003). They are both examples of the nonlinear interactions between vegetation and water availability and they were both introduced as spatially extended interactions. These systems are known to exhibit a variety of spatial arrangements of vegetation resulting from a self-organization process that has been successfully modeled.[1] Some examples of such vegetation patterns

---

[1] Specifically, these systems seem to exhibit a so-called Turing-like instability, where nonlinearities together with spatial processes such as diffusion can generate the observed regular structures.

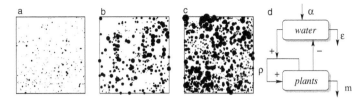

Figure 12.1. Spatial patterns of vegetation cover are known to spontaneously develop under a wide range of conditions in ecosystems. In (a–c) we show three examples of such patterns in (a) arid, (b) semiarid, and (c) Mediterranean ecosystems (adapted rom from Shnerb et al. 2003). These three snapshots involve increasing rainfall levels. The basic interactions between water availability and vegetation cover are summarized in (d).

are shown in figure 12.1a–c for arid, semiarid, and Mediterranean ecosystems, respectively.

Without taking into account this spatial component, we can consider the mean field model just by looking at how water and vegetation cover interact. The basic scheme of plant-water interactions is shown in figure 12.1d. Here w and n indicate water and vegetation cover, respectively. Water is an external input (rain) that we indicate by means of a constant α, and it is depleted by run-off proportional to the current amount $(-\epsilon w)$. Additionally, we need to introduce the consumption of water by vegetation. Such consumption can be linear: the higher the plant density, the larger the water consumption. But it can also be nonlinear, represented here by a quadratic term. Such a term tries to capture underlying plant-plant interactions, in particular possible *facilitation* effects: as more plants become present, better local conditions enhance further growth. Moreover, plants die at some density-independent rate m and grow by exploiting available water with some given efficiency ρ. These basic assumptions allow a two-dimensional dynamical system:

$$\frac{dw}{dt} = \alpha - \epsilon w - w\Gamma(n) \qquad (12.1)$$

$$\frac{dn}{dt} = -mn + \rho w\Gamma(n) \qquad (12.2)$$

where the function Γ introduces the functional form of the water-vegetation interaction. Here the two cases are considered: $\Gamma(n) = n$ (Shnerb et al. 2003) and the quadratic dependence associated to facilitation effects $\Gamma(n) = n^2$ (Klausmeier 1999). The latter can easily be interpreted: the "encounter" of neighboring plants in a given area enhances the likelihood of each to exploit available water resources. For example, a given plant with well-established roots will help other plants to establish themselves.

This model can be further reduced by considering that water dynamics proceeds much faster than vegetation dynamics. On a first approximation we would have $dw/dt \approx 0$ and thus we can replace w in the second equation by

$$w \approx \frac{\alpha}{\epsilon + \Gamma(n)} \qquad (12.3)$$

which gives, once introduced in dn/dt, a one-dimensional system:

$$\frac{dn}{dt} = \frac{\alpha\rho\Gamma(n)}{\epsilon + \Gamma(n)} - mn \qquad (12.4)$$

This general equation contains two different cases, associated to a linear and nonlinear $\Gamma(n)$ term. It also defines two different types of transitions and will be analysed below.

12.3 Continuous Transitions

Let us first consider the linear case. Here the dynamics is described by:

$$\frac{dn}{dt} = \frac{\alpha\rho n}{\epsilon + n} - mn \qquad (12.5)$$

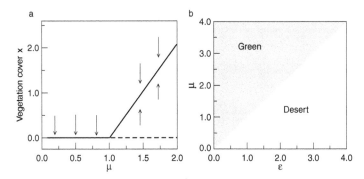

Figure 12.2. Transitions from vegetated to desert states in Shnerb's et al. model. In (a) we plot the bifurcation diagram for $\epsilon = 1$. The corresponding parameter space is displayed in (b) where the two basic domains are indicated, separated by the critical boundary $\mu_c = \epsilon$.

and it is easy to show that two fixed points are possible and correspond to the desert state $n^* = 0$ and the vegetated state

$$n^* = \frac{\rho\alpha}{m} - \epsilon \tag{12.6}$$

The stability of these two points can be determined from $\lambda(0) = \rho\alpha/\epsilon - m$, which gives us a critical condition for the transition from one phase to the other.

Using the notation $\mu \equiv \rho\alpha/m$, we have a desert state for $\mu > \epsilon = \mu_c$ and a vegetated state otherwise. This μ (control) parameter is directly dependent on water availability and plant efficiency, whereas it is inversely proportional to vegetation survival rate. It thus properly incoporates all the relevant trade-offs introduced by the model. In figure 12.2a we show the bifurcation diagram for this system (when $\epsilon = 1$) and the corresponding phase space, where the critical line $\mu_c = \epsilon$ separates both scenarios (figure 12.2b). As we can see, the transition is second-order, with a continuous change of the order parameter n^*.

12.4 Sudden Shifts

The behavior for the second case, when $\Gamma(n) = n^2$, is rather different. In this case, three fixed points are at work. The first is again the trivial one, $n^* = 0$. This fixed point is always stable, since $\lambda_\mu(0) = -1 < 0$, indicating that small vegetation covers will end up in a desert state. The two other fixed points are obtained from the quadratic equation:

$$n^{*2} - \mu n^* + \epsilon = 0 \tag{12.7}$$

and this gives

$$n_\pm^* = \frac{1}{2}[\mu \pm \sqrt{\mu^2 - 4\epsilon}] \tag{12.8}$$

which exist provided that

$$\mu > \mu_c = 2\sqrt{\epsilon}. \tag{12.9}$$

The nontrivial fixed points (when they exist) are separated by a gap from the zero state. Since the latter is always stable, for consistency the two other fixed points, n_-^*, and n_+^*, will be unstable and stable, respectively.[2] The corresponding bifurcation diagram and phase plot appears in figure 12.3a. This diagram reveals something of great importance related to the potential irreversibility of the shift. Starting from a vegetated state, as we change μ (say, water availability) below μ_c, the transition toward desertification cannot be reversed by simply moving back to $\mu > \mu_c$. The system is now locked in the new attractor and no jump back will occur, unless we enter again in to the domain $n > n_-^*$.

We can show the emergence of a bifurcation leading to a three-fixed-point situation by plotting the two components defining

---

[2] Since $n_+^* > n_-^* > n_0^*$, they define an ordered sequence on the real line. The stability of n_0^* implies that trajectories come out from n_-^* and similarly flow into n_+^*, which is thus stable.

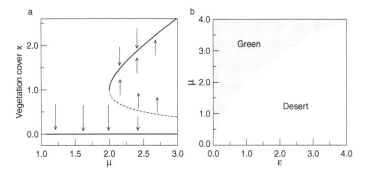

Figure 12.3. Transitions from vegetated (green) to desert states in Klausmeier's model. In (a) we plot the bifurcation diagram for $\epsilon = 1$. The corresponding parameter space is displayed in (b) where the two basic domains are indicated, separated by the critical boundary $\mu_c = 2\sqrt{\epsilon}$.

our reaction terms. To make the analysis simpler, let us fix $\epsilon = 1$ and divide all terms in the differential equation by m. In this way we have a single parameter model:

$$\frac{dn}{dt} = \frac{\mu n^2}{1 + n^2} - n \qquad (12.10)$$

where time has been rescaled by the mortality rate m. Since $n^* = 0$ is already a fixed point, the two functions are reduced to a nonlinear one, namely

$$y_1(n) = \frac{\mu n}{1 + n^2} \qquad (12.11)$$

and a constant one, $y_2(n) = 1$. The nontrivial fixed points are reached at the intersection $y_1(n) = y_2(n)$, which means the crossings between the horizontal line $y = 1$ and the curve y_2.

A simple analysis reveals that the critical value $\mu_c = 2$ is indeed the bifurcation point of our system (figure 12.4a). The curve y_2 is such that $y_2(0) = 0$, but $y_2 \to 0$ for $n \to \infty$. Since it is positive for the $n > 0$ meaningful interval, we should expect to find a maximum at some given value n_m. The extrema of $y_2(n)$ are

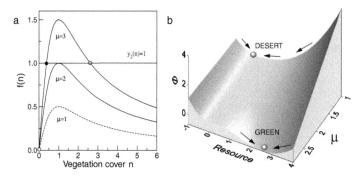

Figure 12.4. The presence and character of the fixed points in Klausmeier's model can be determined by using a simple graphical method (a) where the two components of the dynamics are plotted (see text). In (b) we show the predicted potential function $\Phi_\mu(n)$ for different values of the bifurcation parameter μ. As we cross $\mu_c = 1$, the equilibrium represented by the marble shifts from a nonzero (green) vegetation to the desert state.

obtained from the condition $dy_2/dn = 0$, which gives $n_m = 1$. Now we can calculate the value of the maximum y_m taken by our function at this point. We obtain

$$y_2(n_m) = \frac{\mu}{2} \qquad (12.12)$$

and this immediately allows us to determine the conditions for crossing with $y_1 = 1$: if $\mu < 2$ no crossing is possible since the curve is below $y = 1$ (figure 12.4a). Once we cross this value, two intersections are expected to happen.

Finally, we can also calculate the form of the potential function $\Phi_\mu(n)$ whose minima are associated with desert or vegetated states. By using Klausmeier's model, we obtain:

$$\Phi_\mu(n) = \frac{n^2}{2} - \mu \, (n - \arctan n) \qquad (12.13)$$

which is depicted in figure 12.4b for different values of μ. As predicted by our previous stability analysis, the marble at the bottom of $\Phi_\mu(n)$ will move from $n^* > 0$ to the desert state $n^* = 0$ as we cross the critical level μ_c.

12.5 Metapopulation Dynamics

Let us now consider the problem of species extinctions under habitat loss. The simplest, first approximation to this problem is obtained by exploring the consequences of habitat loss in a metapopulation (Hanski 1999; Bascompte and Solé 1996). A metapopulation can be defined as a set of geographically distinct local populations maintained by a dynamical balance between colonization and extinction events. Such events can be described in terms of state transitions happening with a given probability. Let us start this section by revisiting the Levin's (1969) model, which captures the global dynamics of a metapopulation.

The basic model reads:

$$\frac{dx}{dt} = cx(1 - x) - ex \qquad (12.14)$$

where x is the fraction of patches occupied, and c and e are the colonization and extinction rates, respectively. This model has an extinction state $x_0^* = 0$ and a nontrivial solution given by Levin's steady population, $x_L^* = 1 - e/c$. The colonization rate has to be larger than the extinction rate for the metapopulation to persist. The resulting model has the same structure as the one already seen in chapter 9 on epidemics. It can also be rewritten as a logistic-like model as follows:

$$\frac{dx}{dt} = (c - e)x \left(1 - \frac{x}{x_L^*} \right) \qquad (12.15)$$

where the condition for stability of x_L^* becomes obvious since we need $c > e$ in order to have a positive growth rate.

One can easily introduce habitat loss into the framework of model (12.14). If a fraction D of sites is permanently destroyed,

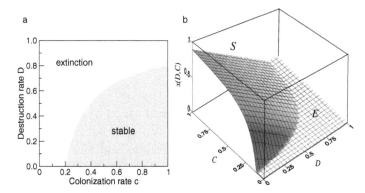

Figure 12.5. Phase changes in metapopulations under habitat loss. In (a) the two phases of Levin's model, namely extinction (white area) and stable (gray) are shown. In (b) the corresponding order parameter $x^*(c, D)$ is shown (for $\epsilon = 0.1$.) Either if colonization is low or habitat destruction too high, extinction occurs by crossing the critical boundary $D_c = 1 - e/c$.

this reduces the fraction of vacant sites that can potentially be occupied. Model (12.14) becomes now:

$$\frac{dx}{dt} = cx(1 - D - x) - ex. \qquad (12.16)$$

This model has two equilibrium points (to be obtained from $dx/dt = 0$): $x^* = 0$ (extinct population) and $x^* = 1 - D - e/c$. The latter decays linearly with habitat loss, becoming zero when $1 - D - e/c = 0$. This condition gives a critical destruction level $D_c = 1 - e/c$ indicating that a nontrivial dependence exists between available patches and the species-specific extinction and colonization properties. Once we cross this threshold (i.e., if $D > D_c$) the population goes inevitably extinct. This situation is illustrated in figure 12.5. Here the critical line separates the two qualitative types of behavior. As we approach the critical value D_c by increasing the amount of habitat destroyed, the frequency of populated patches decays linearly, becoming zero at the boundary.

This process is actually accelerated by considering local dispersal (Bascompte and Solé 1996).

The previous models can be generalized in order to observe first-order transitions by incorporating an additional constraint. It explains a common observation from the analysis of field data: it appears that species occur, at any one time, either in most potential sites or only in a few. This bimodal pattern can be easily explained by means of the so-called Allee effect, which introduces a negative contribution to population growth at low densities (Stephens and Sutherland 1999; Fowler and Ruston 2002; Liebhod and Bascompte 2003; Courchamp et al. 2009). This can be for a number of reasons, from mate shortage, failure to exploit available resources, to lack of (required) cooperation. It can also be induced by anthropogenetic means (Courchamp et al. 2006).

The Allee effect can be easily incorporated within Levin's formulation by assuming that there is some additional constraint limiting the survival of local populations. This can be done by properly modifying the model (Dennis 1989; Amarasekare 1998; Keitt et al. 2001; Courchamp et al. 2009):

$$\frac{dx}{dt} = \frac{(c-e)}{x_L^*}x(x-a)\left(1 - \frac{x}{x_L^*}\right) \qquad (12.17)$$

Here a new parameter a, such that

$$0 < a < 1 - \frac{e}{c} \qquad (12.18)$$

and defining a threshold fraction of occupied sites below which populations go extinct. The new model involves a cubic equation, and thus three fixed points are present. Two of them are identical to those found in the original model, whereas the third, $x^* = a$, is new. It introduces an intermediate state that strongly changes the structure of the bifurcation diagram.

It is easy to show now that the extinction point $x^* = 0$ is always stable, in sharp contrast with the original Levin's model, where

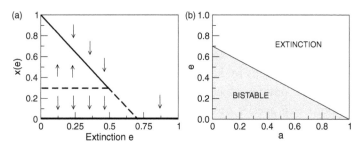

Figure 12.6. Levin's model with multiple states due to the Allee effect. Here we use $a = 0.3$ and $c = 0.7$ and plot the bifurcation diagram (a) and the paramater space (b) for different values of extinction. As we can see in (a) a bistable state is generated, the $x^* = a$ fixed point the unstable one.

this state depends on the parameters. By studying the other two fixed points we can see that the x_L^* is now stable if $c(1 - a) < e$ whereas the second is unstable (it does not exist outside this domain). The resulting bifurcation diagram is shown in figure 12.6a. We can see that it describes a first-order transition: once we cross the $e_c = c(1 - a)$ critical boundary, the population shifts to zero. The model can be further explored (Amarasekare 1998) by incorporating the destruction rate:

$$\frac{dx}{dt} = (c(1 - D - x) - e)x \left(\frac{c(x - a)}{c(1 - D) - e} \right) \quad (12.19)$$

We leave to the reader the analysis of this extended model. Let us just mention that metapopulation persistence now requires more strict conditions.

12.6 Warning Signals

If critical changes can generate catastrophic events by slowly tuning a key parameter around the transition point, it would be desirable to have warning signals indicating that such changes are likely to occur (Scheffer 2009; Scheffer et al. 2009). Some of such early warning signals have been shown to exist. They are associated to both time- and spatial-dependent ingredients. Specifically, it has been shown that critical slowing down (see section 3.4) can be expected when approaching the threshold (Chrisholm and Filotas 2009). Here natural fluctuations would be amplified by the internal nonlinear dynamics and a high variance of population change should observable. Similarly, some special types of spatial patterns are known to arise close to the transition (Riekert et al. 2004; Kefi et al. 2007; Kefi 2009). As it occurs with magnetic systems (chapter 1), spatial correlations experience dramatic changes close to criticality. It has been reported from both ecological and climate studies that marked trends in variance measures typically occur close to criticality (Lenton 2008; Anderson 2009; Dakos et al. 2008; van Nes and Scheffer 2007).

A special problem, addressed in terms of phase transitions in ecology, concerns the propagation of fires. Forest fires are known to be an active component of ecosystem dynamics and are an important evolutionary force (Bond and Keeley 2005). They promote and maintain diversity by creating a range of opportunities in local habitats of varying conditions. Fires are actually similar to herbivores (Bond and Keeley 2005), and, as mentioned in chapter 9, their dynamical behavior is closely related to that exhibited by epidemic spreading. Although fires are a natural outcome of ecosystem dynamics, their frequency and size are growing across the tropics. Their interactions with increased habitat degradation, logging, and climate change have been responsible for huge carbon emissions (Cochrane 2003). By analyzing the patterns displayed by Amazonian forest fires, it is possible to predict some potential scenarios of megafire spreading (Pueyo 2007; Pueyo et al. 2010). These studies suggest that, if the most adverse forecasts are realized, the second-order transition expected from an epidemics-like behavior would be replaced by sudden (first-order) changes.

13

INFORMATION AND TRAFFIC JAMS

13.1 Internet and Computer Networks

One seemingly general property of transport networks, including roads, Internet, the power grid, or a highway, is the undesired emergence of jamming and congestion. As the flow of matter, energy, and information increase beyond a given threshold, problems start to pile up. This situation is common in roads when many vehicles become jammed, but it is also a problem for the power grid when energy demands exceed the grid's capacity. Congestion is actually one of the oldest examples of a phase transition that we can think about. Who hasn't experienced a traffic jam or an Internet storm? But few people realize that, both in highways and computer networks, it involves a phase transition between fluid behavior and congestion (Nagel and Schreckenger 1992; Nagel and Rasmussen 1994; Huberman and Lukose 1997; Ohira and Sawatari 1998; Solé and Valverde 2000). The presence of two phases has been reported in different field studies. An example is shown in figure 13.1, where the fluctuations of packets flowing through a given link in a computer network are displayed. Over some intervals, the link becomes congested by high traffic, which saturates it, whereas for other intervals a lower density fluctuation is seem (Takayasu et al. 2000).

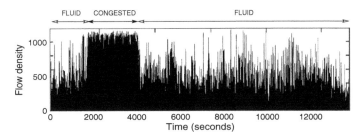

Figure 13.1. An example of time sequence of flow density fluctuations observed at intervals of 0.1 s in an Internet node (redrawn from Takayasu et al. 2000). An intermediate interval reveals a very high flow of traffic, making the system highly congested. The other two intervals reveal a highly fluctuating, but otherwise fluid, traffic regime.

Parallel multiprocessor networks have been shown to display complex dynamics and a phase transition separating a congested from a noncongested phase, both in the real and the simulated (Ohira and Sawatari 1998; Solé and Valverde 2000) counterparts.

In real systems, there is a coupling between the system's dynamics (exhibiting fluctuations) and the human agents, which both create and respond to such changes. This feedback is clear if we take into account how users interact within the Internet while searching for information. Each user sends and receives information and as a consequence contributes to the traffic. However, when the flow becomes too slow because of a congestion event, the behavior of the users changes: congestion reduces their pressure on the system and eventually a more fluid state is recovered. Such feedback has been introduced in models of Internet traffic (Valverde and Solé 2002, 2004), which have been shown to consistently reproduce several key properties of real computer traffic, suggesting that the complex fluctuations might result from a critical state.

In its simplest form (to be considered here, see figure 13.2a), we can imagine a computer network as a set of connected elements, some of which (the so-called hosts) create and send packets to some other chosen host node. The other type of element, the so-called routers only canalize the traffic through the web and cannot store packets. Instead, hosts can store arriving packets in a queue, which is constantly shortened as the host processes the incoming messages (which are sequentially piled following their order of arrival). Once the packet is received, it is either removed (if the queue is empty) or appended at the tail of the host queue. These hosts are assumed to be randomly scattered over the lattice. Such a pattern matches some special types of well-known computer architectures (Hillis 1984) and has been explicitly implemented in a grid structure (Bolding et al. 1997). At low rates of packet injection, we should expect fluid traffic, whereas for high rates, the system will inevitably become congested. Below we develop the simplest mean field approach for this problem, showing that two phases separated by a sharp transition should occur.

13.2 Mean Field Model of Traffic Flow

Although the grid shown in figure 13.2a implies spatial correlations between nodes, we ignore them and assume that the nodes are somewhat mixed, displaying an average number of connections $\langle k \rangle$ among them. We also consider queues having a potentially unlimited length. A simple mean field model can be obtained for the total density of packets, namely $\Gamma(t) \equiv N(t)/L^2$.

The number of traveling packets increases as a consequence of the constant pumping from the hosts, which are present at a constant density ρ. If each host sends packets at a rate μ the total rate of packet emission is $\rho\lambda$. However, packets are efficiently removed from the system if the lattice is not too congested (i.e., if free space for movement is available) but accumulate as a

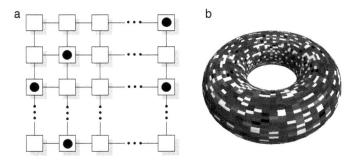

Figure 13.2. A simplified model of computer traffic. Here (a) the elements are arranged on a square grid with $L \times L$ nodes. Two types of nodes are considered: hosts (indicated with filled circles) and routers (plain squares). Hosts send and receive packets, whereas routers simply redirect them through the grid. The grid is folded in such a way that we form a torus (b). In a model of computer traffic, different nodes receive different amounts of traffic load, and different levels of congestion will be observed. In (b) this is indicated by lighter gray values.

consequence of already jammed nodes. We can roughly state that once the number of packets exceeds the number of lattice sites, congestion will lead to packet accumulation. Since the total lattice size is L^2, we can consider this value the critical limit of packet load that the system can manage without getting congested.

The time evolution of the density of packets will follow the mean field equation for the number of packets N:

$$\frac{dN}{dt} = \rho \mu L^2 - \alpha \langle k \rangle N \left(1 - \frac{N}{L^2} \right) \qquad (13.1)$$

As we can see here, we adopt a basic criterion for congestion: if the number of packets in the system is smaller than the grid size ($N < L^2$), no congestion takes place, whereas if $N > L^2$, under our mean field picture, it is likely that congested traffic will be present.

Using the previous definitions the model can be normalized and it reads:

$$\frac{d\Gamma}{dt} = \rho\mu - \alpha\langle k\rangle\Gamma\,(1-\Gamma) \tag{13.2}$$

The last term on the right-hand side indicates the rate of removal, which is proportional to the number of input pathways (number of neighbors $\langle k\rangle$) available to incoming packets and the efficiency α with which they are removed. The rate α corresponds to the inverse of the average lifetime of packets (Fuks and Lawniczak 1999).

13.3 Fluid Flow and Congestion Explosion

In our approximation, we consider a fixed, constant injection of packets from the hosts. For λ constant (i.e., when no reactive responses are assumed to exist) we have the two fixed points, namely:

$$\Gamma_\pm = \frac{1}{2}\left[1 \pm \sqrt{1-\frac{4\rho\mu}{\alpha\langle k\rangle}}\right] \tag{13.3}$$

These two fixed points exist provided that $\mu < \mu_c$ where

$$\mu_c = \frac{\alpha\langle k\rangle}{4\rho} \tag{13.4}$$

The stability of these two fixed points can be determined from the sign of $\lambda_\mu = -\alpha\langle k\rangle\,[1-2\Gamma^*]$, which in our case reads

$$\lambda_\mu(\Gamma_\pm) = \alpha\langle k\rangle\left[1 \pm 2\sqrt{1-\frac{\mu}{\mu_c}}\right] \tag{13.5}$$

and thus we conclude that Γ_+ is unstable whereas the lower branch Γ_- is stable. Outside this parameter domain, the system undergoes an explosive growth of congestion.

The corresponding bifurcation diagram is shown in figure 13.3a. This is a rather special diagram, involving a single

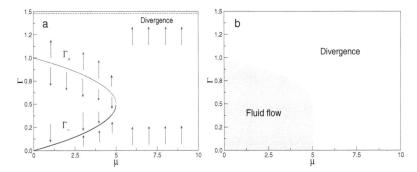

Figure 13.3. (a) Bifurcation diagram for the traffic congestion model. Here the bifurcation parameter is the rate of packet emission $(\mu\rho)$ by each element in the grid (we fixed $\langle k \rangle = 2$ and $\rho = 0.1$). For $\mu < \mu_c$ two fixed points are present but only one of them (the lower branch) is stable. After the critical emission rate is crossed, the system diverges and congestion increases indefinitely, leading to the system's collapse. In (b) an alternate representation is given, showing the two basins of attraction defined by the system dynamics. The gray zone involves all the initial values that lead to the stable, fluid state.

equilibrium state toward which a range of initial conditions converges. The two branches obtained from our analysis are indicated for $\mu < \mu_c$ and we can see that the diagram is similar to other plots associated with first-order transitions. But here the lower branch is not the unstable one: the order has been exchanged, meaning that initial conditions starting from values $\Gamma(0) > \Gamma_+$ will lead to increasing congestion. Similarly, the lack of a (finite) fixed point for $\mu > \mu_c$ needs to be interpreted. But the solution is simple: once we have enough packets in our system, a runaway effect generates an explosive increase that creates a density of packets crossing the $\Gamma = 1$ value. Afterward, the last term in the mean field differential equation will turn positive and congestion will allways rise. The two-phase behavior is illustrated by the trajectories following the system, as shown in figure 13.4a. The corresponding phases are also shown in figure 13.4b, separated by the critical line $\mu_c = \alpha \langle k \rangle / 4\rho$.

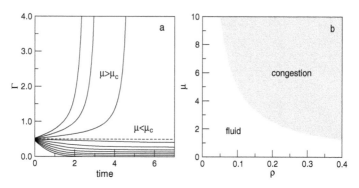

Figure 13.4. Phase transition in the mean field model of traffic. In (a) we show the trajectories obtained numerically using as initial condition $\Gamma(0) = 0.5$ and different values of μ (other parameters are $\langle k \rangle = 2$ and $\rho = 0.1$), which gives a critical rate $\mu_c = 5$. For μ values below critical, the system's activity relaxes to a stable value Γ_-^* (see text). In (b) we show the phase space for this model using μ and ρ as control parameters. The shadowed area corresponds to a divergent number of packets.

13.4 Self-Regulated Traffic

In standard computer networks, congestion levels are bounded for two good reasons. First, queues are not infinite and thus unbounded growth of stored packets cannot occur. Second, and most important, the rate of packet emissions cannot be independent of congestion levels. The feedback between the two components can trigger complex fluctuations (not taken into account here) and scaling behavior (Huberman and Lukose 1997; Valverde and Solé 2002, 2004). In this context, it has been shown that model computer networks spontaneously evolve to an intermediate state close to the critical boundary separating fluid traffic from congestion. Such a situation is known as self-organized critical and has been suggested to be present in many different situations both in nature and society (Bak 1997).

Under our mean field approximation, a simple (but important) modification allows us to show that such a critical state is achieved. Considering that the level of packet occupation $1 - \Gamma$ affects not only packet dynamics but also rate of emission, we can indicate a tentative model as follows:

$$\frac{d\Gamma}{dt} = (\rho\mu - \alpha\langle k\rangle\Gamma)(1 - \Gamma) \qquad (13.6)$$

which now defines two stable equilibrium points, namely $\Gamma = 1$ and $\Gamma = \rho\mu/\alpha$. We can easily check that

$$\left(\frac{\partial\dot{\Gamma}}{\partial\Gamma}\right)_{\Gamma=1} = -\rho\mu + \alpha\langle k\rangle = -\left(\frac{\partial\dot{\Gamma}}{\partial\Gamma}\right)_{\Gamma=\rho\mu/\alpha} \qquad (13.7)$$

and thus a nonsaturated point will be stable provided that $\mu < \mu_c = \alpha\langle k\rangle/\rho$. Otherwise, congestion rises until complete collapse occurs. This result is consistent with a continuous phase transition, and the critical point is clearly similar (up to some constant factor) to the one just discussed above. Now however, for a fixed $\mu < \mu_c$, a single solution exists, where the system achieves a steady state with intermediate levels of traffic.

As we have mentioned in other situations, the architecture of the actual web of computer interactions will also have an impact on its behavior. The Internet is known to be a complex heterogeneous network, with most nodes having only a few connections whereas a handful of nodes display a huge number of links (Yook et al. 2002). Such a topological pattern makes these webs very efficient but fragile, and the strategies of packet routing play a relevant role in making traffic efficient. In this context, other types of phase transitions can be found. An interesting example concerns the depth of routing tables (Valverde and Solé 2004). As we outlined above in our lattice model, routers redirect packets toward their destination. However, in a multimillion-site web the routing tables are limited to a maximum depth. Beyond this horizon, the packet would travel randomly. Although it might appear optimal to have huge routing tables spanning the whole network, it can actually be shown that there is a critical path horizon beyond which packets are successfully delivered.[1] This critical horizon corresponds to a phase transition (Valverde and Solé 2004) and shows that an appropriate combination of routing and randomness leads to an optimal information flow at a low cost.

---

[1] In fact, a system incorporating full tables will collapse, since some particular links will be typically chosen as key elements within the shortest paths and experience traffic jams.

14

COLLECTIVE INTELLIGENCE

14.1 Swarm Behavior

One of the most remarkable events in the evolution of life was the emergence of social behavior among insects. The generation of complex societies of cooperating individuals represented a major leap in the history of our biosphere. It is difficult not to become fascinated by the plethora of patterns displayed by the collective work of ants, termites, bees, and social wasps. The huge nests created by termites or the army ant raid patterns traveling through the rainforest are just two examples (Holldobler and Wilson 2009). The ecological relevance of these insects is illustrated by the following fact: in some rainforests, the dry weight of ants and termites is about four times that of all the other land animals (Holldobler and Wilson 1990).

Social insects build complex structures, scan their environments in search of available resources, perform group-based decisions (Deneubourg and Goss 1989; Camazine et al. 2002), and—in many ways—remind us how brains work (Hofstadter 1980; Solé and Goodwin 2001). In social insects, while colonies behave in complex ways, the capacities of individuals are relatively limited. But then, how do social insects reach such remarkable ends? The answer comes to a large extent from self-organization:

Figure 14.1. Symmetry-breaking in ant colonies. Here two examples of experiments, illustrating how the collective behavior of the colony leads to a broken symmetry. In (a) escaping ants choose preferentially one of the two possible exits (symmetrically located on both sides of a circular arena). The image in (b) shows ants moving through one of two bridges between the nest and a food source. After a while, all use one path (picture kindly provided by Ernesto Althsuler [a] and by Guy Theraulaz[b]).

insect societies share basic dynamic properties with other complex systems (Millonas 1993; Gordon 1999, 2010; Detrain and Deneubourg 2006; Sumpter 2006) and exploit bifurcations and nonlinearities as their source of internal organization and decision making (Deneubourg et al. 1989; Bonabeau 1996). Moreover, some trends displayed by the global organization of insect societies, such as the diversity of morphological casts, nest organization, and hierarchy have been related to phase transition phenomena (Oster and Wilson 1978; Valverde et al. 2009).

Several experiments clearly indicate that amplification of perturbations by means of individual interactions leads to collective responses. Two examples are shown in figure 14.1. In the first (a) a small group of ants in a circular arena suddenly experience a panic reaction to an externally introduced chemical signal. The arena is enclosed by a plastic tube with two identical exists, but ants tend to choose one of them: a symmetry-breaking phenomenon pervades the panic behavioral response. The second example, to be analyzed here (b) deals with how ant colonies can choose among two given bridges connecting their nest with a food source.

Foragers follow the chemical trails left by their fellow ants. The more pheromone is deposited on a given path, the higher the probability that ants follow that path and leave further chemical marks. When the paths have different lengths, ants typically end up choosing the shortest, thus solving an optimization problem (Deneubourg et al. 1983, 1987; Millonas 1992; Bonabeau 1996; Beckers et al. 1993; Deneubourg et al. 1990). But even when the two bridges are identical, the amplification process can lead to a choice, thus breaking the symmetry of the system. Here we will consider this special case.

In this chapter we study two different types of phase transition phenomena that have been reported from the experimental analysis of social insects. They illustrate how collective behavior can allow colonies to make decisions and display self-organization behavior.

14.2 Foraging and Broken Symmetry

The two-bridges problem can be formalized as follows. Let us indicate by x_1 and x_2 the concentrations of trail pheromone in each branch. These concentrations will be somewhat proportional to the number of ants exploring each path. The equations for these quantities will read (Nicolis and Deneubourg 1999):

$$\frac{dx_1}{dt} = \mu q_1 P_1(x_1, x_2) - \nu x_1 \qquad \frac{dx_2}{dt} = \mu q_2 P_2(x_1, x_2) - \nu x_2 \tag{14.1}$$

where μ is the rate of ants entering each branch from the nest, q_i the rate of pheromone deposition in the $i - th$ branch, and ν the rate of pheromone evaporation. The functions $P_i(x_1, x_2)$ introduce the choices made by ants (which bridge is chosen) given the pheromone concentration. A possible response function is described by (Beckers et al. 1992; Deneubourg et al. 1990):

$$P_i(x_1, x_2) = \frac{(x_i + K)^2}{\Theta(x_1, x_2)} \tag{14.2}$$

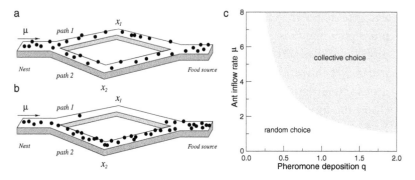

Figure 14.2. Symmetry breaking in the food search experiment. When two identical bridges are used, the amplification process associated to pheromone fields forces the collective to make a choice, breaking the initial equivalence.

where $\Theta(x_1, x_2) = \sum_{j=1,2}(x_j + K)^2$ and $i = 1, 2$. It can be shown that this model allows interpretation of the minimization process in terms of a symmetry-breaking bifurcation.

Let us consider first the symmetric case where $q_1 = q_2 = q$ and thus the set of equations:

$$\frac{dx_1}{dt} = \mu q \frac{(x_1 + K)^2}{\Theta(x_1, x_2)} - \nu x_1 \qquad \frac{dx_2}{dt} = \mu q \frac{(x_2 + K)^2}{\Theta(x_1, x_2)} - \nu x_2$$

$$(14.3)$$

This symmetric scenario is not very interesting in terms of decision making, since there is no better solution. But the exercice is useful and shows that a choice can be made by amplification of fluctuations. The same principle pervades the more general case, to be explored later. It is not difficult to show that one possible solution to this system is the symmetric fixed point $x_1^* = x_2^* = x^*$ when ants equally distribute themselves over both branches. For this special case, we have $P_i(x_1, x_2) = P(x) = \mu q/2$ and thus a single equation $dx/dt = \mu q/2 - \nu x$ that gives a fixed point $x^* = \mu q/(2\nu)$. The second state would correspond to the choice of one of the branches. Since the total pheromone concentration is $2x^* = \mu q / v$, we can see that

$$\left(\frac{\mu q}{v} - x \right)(x + K)^2 = x \left(\frac{\mu q}{v} - x + K \right)^2 \qquad (14.4)$$

After some algebra, this gives the new fixed points $x_{\pm}^* = (x_1^*, x_2^*)$ with

$$x_1^* = \frac{\mu q}{2v} + \sqrt{\left(\frac{\mu q}{2v} \right)^2 - K^2} \quad x_2^* = \frac{\mu q}{2v} - \sqrt{\left(\frac{\mu q}{2v} \right)^2 - K^2} \qquad (14.5)$$

and $x_i^* = \mu q / v - x_j^*$, $(i \neq j)$. This pair of fixed points will exist provided that $\mu q / 2v > K$, which allows us to obtain the critical point at

$$\mu_c = \frac{2Kv}{q} \qquad (14.6)$$

Below this value, the only fixed point is the symmetric case with identical flows of ants in each branch. This critical curve allows to define two phases, as indicated in figure 14.2c, where we plot $\mu_c = \mu_c(q)$ using $K = v = 1$. The gray area marks the phase where collective choices are possible, whereas the white area is associated with random behavior and no amplification of fluctuations. In this latter domain, no symmetry-breaking is possible. Since the symmetric point x^* becomes unstable and we have two new symmetrically located branches around it, these two new solutions must be stable, provided that $\mu > \mu_c$. The corresponding bifurcation diagram is shown in figure 14.3. This symmetric model can be generalized to (more interesting) asymmetric scenarios where the two potential choices are different (see Detrain and Deneubourg 2009 and references cited) either because the food sources have different size or because paths have different lengths and the shortest path must be chosen.

Transitions in colony behavior associated to food search have been characterized in other contexts, sometimes displaying first-order transitions (Beckman et al. 2001; Loengarov and Tedeshko 2008).

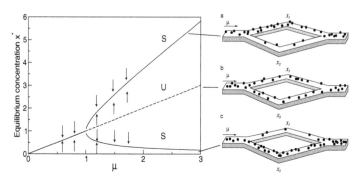

Figure 14.3. Bifurcations in the symmetric bridge model of collective choice. Here we have used $K = 1$, $q = 2$, $v = 1$ and tuned the parameter μ (here $\mu_c = 1$). The straight line gives the symmetric state $x^* = \mu q / 2v$. The two branches define the two alternative states where one branch has been chosen. The corresponding patterns of ants walking on each branch are indicated at right.

14.3 Order-Disorder Transitions in Ant Colonies

In this section we will address a different type of dynamical pattern exhibited by ant colonies, namely the presence of global activity oscillations (Franks and Bryant 1987; Cole 1991a). This type of phenomenon has been described in different genera of ants, but has been particularly well studied using colonies of the genus *Leptothorax*. Briefly, these small colonies (composed of just fifty to one hundred individuals) show a remarkable time-dependent pattern of activity, where periods of no movement are followed by bursts of all-active individuals. These bursts are the outcome of local interactions among ants and appear at almost periodic intervals over time (figure 14.4).

By using groups of ants of different sizes, a series of experiments revealed that the density ρ of ants enclosed within a given arena had a great impact on the observed dynamics. Specifically, it was found that the pattern of synchronization exhibited by the colony was *not* present at the individual level. In other words, the

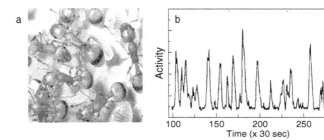

Figure 14.4. Fluctuations in colony activity are known to occur in many species of ants, such as those of the genus *Leptothorax*. These ants (a) form small colonies (picture by Bernhard Seifert) exhibiting almost regular bursts of global activity (b) characterized by intervals of little activity followed by intervals where most or all ants are performing some task (adapted from Cole 1991a).

global clock was not the result of the synchronization of multiple individual clocks, as it occurs in other situations (Strogatz 2003). Coherent bursts of activity appear close to a given critical density ρ_c but there is no regular behavior below the threshold. Instead, individual ants appear to display chaotic behavior (Cole 1991b). The coherent pattern displayed by the colony as a whole has been explained in terms of nonlinear dynamics (Goss and Deneubourg 1988; Solé et al. 1993; Cole 1993; Boi et al. 1998; Delgado and Solé 1997, 2000). In an inactive colony, an individual can become spontaneously active, moving around and exciting neighboring ants, which also move and can activate others. A percolation-like phenomenon takes place at the critical density.

The simplest mean field model that can be formulated is closely related to the one obtained in chapter 9 for the propagation of epidemics. As suggested in Cole (1991c), the propagation of activity in ant colonies is similar to the epidemic-spreading model. Let us now consider our population composed of two groups, namely active and inactive ants, whose numbers will be indicated by A and I, respectively. Using the normalized values $x = A/N$ and $y = I/N$, and assuming that the density of occupied space is $\rho \in [0, 1]$, the mean field equation for the fraction of active ants is:

$$\frac{dx}{dt} = \alpha x(\rho - x) - \gamma x \qquad (14.7)$$

An important difference in relation to epidemic spreading is that the density of ants is now a key control parameter.

The equation displays two equilibrium points. The first corresponds to no activity at all ($x_0^* = 0$) while the second is associated to the presence of a given level of activity, namely

$$x_1^* = \rho - \frac{\gamma}{\alpha} \qquad (14.8)$$

The colony will display nonzero activity (i.e., x_1^* will be stable) provided that

$$\lambda_\mu(0) = \alpha\rho - \gamma > 0 \qquad (14.9)$$

which, in terms of the density of ants corresponds to the condition $\rho > \rho_c = \gamma/\alpha$.

The deterministic model predicts that no activity will be present at subcritical densities, whereas experiments indicate that activity is present, largely due to random individual activations, at small densities. Moreover, it has been found that fluctuations increase close to the critical point, consistent with real data (Solé et al. 1993; Miramontes 1995). This is illustrated by the shape of the associated potential, which now reads:

$$\Phi_\rho(x) = \alpha\frac{x^3}{3} - (\alpha\rho - \gamma)\frac{x^2}{2} \qquad (14.10)$$

Three examples of this potential are shown in figure 14.5c for subcritical, near-critical, and supercritical cases. As expected, the potential flattens as criticality is approached.

The observed variance in colony fluctuations is a consequence of critical slowing down (see chapter 1), and for this model it can

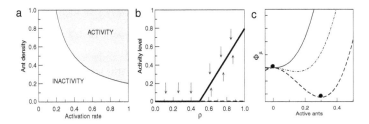

Figure 14.5. (a) The two phases displayed by the mean field model on the (α, ρ) paramater space. The gray area corresponds to the active colony state, where colony activity $x^* > 0$ is present. In (b) the corresponding bifurcation diagram is shown (in both cases we fix $\gamma = 0.2$). Three examples of the associated potential are shown in (c) for $\rho = 0.3, 0.55$ and $\rho = 0.7$ (from top to bottom).

be shown that the relaxation time $\tau(\rho)$, scales as

$$T(\rho) = \frac{1}{\alpha} \int_{x(0)}^{x(T(\rho))} \frac{dx}{x(\rho - \rho_c - x)} \sim (\rho - \rho_c)^{-1} \quad (14.11)$$

As we approach criticality, random activations are likely to propagate for a while before global activation dies out, thus providing the source for higher activity levels and broader fluctuations.

It seems strange that ants display global colony oscillations. A simpler pattern of activity, where all ants are allways active at some low level, would appear more natural. In both cases, the same global average activity could be obtained. There is, however, a good reason for the oscillations. A detailed study of the two alternatives (Delgado and Solé 2000) shows that, functionally, a fluctuating system actually outperforms the alternative (not observed) constant-activity scenario.

14.4 Structure and Evolution in Social Insects

We have presented examples illustrating two ways of solving an optimization problem based on collective behavior. However, insect colonies display multiple levels of complexity, and the presence of critical thresholds is widespread. Thresholds in behavioral responses have been identified in the transition to social behavior, the distribution of workers into castes, the emergence of hierarchical organization in wasps, and aggregation phenomena. Moreover, some key traits exhibited by insect societies seem closely related to phase transitions. A wide range of examples have been analyzed, from worker specialization (Holldobler and Wilson 1990) to nest architecture (Valverde et al. 2009). In the latter case, it has been shown that the spatial structures associated with the organization of termite nests seem to be compatible with a percolation transition.

15

LANGUAGE

15.1 Lexical Change

Human language stands as one of the most important leaps in evolution (Bickerton 1990; Deacon 1997; Maynard Smith and Szathmáry 1995). It is one of evolution's most recent inventions, and might have appeared as recently as fifty thousand years ago. Our society emerges, to a large extent, from the cultural evolution allowed by our symbolic minds. Words constitute the substrate of our communication systems, and the combinatorial nature of language[1] (with an infinite universe of sentences) allows us to describe and eventually manipulate our world. By means of a fully developed communication system, human societies have been able to store astronomic amounts of information far beyond the limits imposed by purely biological constraints. As individuals sharing our knowledge and the cumulative experience of past generations, we are able to forecast the future and adapt in ways that only cultural evolution can permit.

The faculty of language makes us different from any other species (Hauser et al. 2002). The differences between animal

---

[1] More important, language involves recursivity, by which very complex, coherent sentences can be constructed in a nested manner.

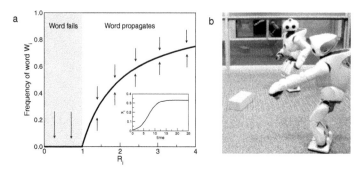

Figure 15.1. (a) Bifurcations in word learning dynamics. If the rate of word learning exceeds one (i. e. $R_i > 1$), a stable fraction of the population will use it. If not, then a well-defined threshold is found leading to word extinction. The inset shows an example of the logistic (S-shaped) growth curve for $R_i = 1.5$ and $x_i(0) = 0.01$. Lexical diffusion also occurs among artificial agents (b) where words are generated, communicated and eventually shared by artificial, embodied agents such as robots (picture courtesy of Luc Steels).

communication and human language are fundamental, both in their structure and function. Although evolutionary precursors exist, it is remarkable to see that there seems to be no intermediate stage between them (Ujhelyi 1996). Such a shift might have occurred for a number of reasons (Wray 2002) from rare events (making human language a rather unique, unlikely phenomenon) to so-called macromutations. But alternative scenarios support the idea that language evolved from simple to complex communication involving error limits (Nowak and Krakauer 1999) or phase transitions (Ferrer and Solé 2003).

Far from being a static structure, language is always changing (McWorter 2003). Even at short time scales of only hundreds of years, as can be seen from the study of written texts (figure 15.1a), languages change fast. Languages themselves appear and disappear (Crystal 2000) and in some ways they behave as living species. As it happens with the fossil record of life, most languages of the world went extinct a long time ago. As John McWorter points out: "Like animals and plants, languages change, split into subvarieties, hybridize, revivify, evolve functionless features and can be even genetically altered." Perhaps not surprisingly, these similarities suggest that it makes sense to think of language and its changes in ecological and evolutionary terms. Many different aspects of language change have been addressed by theoreticians, and evidence indicates that phase transition phenomena might actually contribute to how languages are shaped through time. Here we consider such transitions in two very different contexts.

15.2 Words: Birth and Death

The potential set of words used by a community is listed in dictionaries (Miller 1991), which capture a given time snapshot of the available vocabulary; but in reality speakers use only part of the possible words: many are technical and thus only used by a given group, and many are seldom used. Many words are actually extinct, since no one is using them. On the other hand, it is also true that dictionaries do not include all words used by a community and also that new words are likely to be created constantly within populations, their origins sometimes recorded. Many such words demonstrate new uses of previous words or recombinations, and sometimes they come from technology. One of the challenges of current theories of language dynamics is understanding how words originate, change, and spread within and between populations, eventually becoming fixed or extinct. In this context, the appearance of a new word has been compared to a mutation.

As occurs with mutational events in standard population genetics, new words or sounds can disappear, randomly fluctuate, become or fixed. In this context, the idea that words, grammatical constructions, and sounds can spread through a given population was originally formulated by William Wang to propose and explain how lexical diffusion (i.e., the spread across the lexicon) occurs. Such a process requires the diffusion of the innovation from speaker to speaker (Wang and Minett 2005).

A very first modeling approximation to lexical diffusion in populations should account for the spread of words as a consequence of learning processes (Shen 1997; Wang et al. 2004; Wang and Minett 2005). Such a model should establish the conditions favoring word fixation. As a first approximation, let us assume that each item is incorporated independently (Shen 1997; Nowak et al. 1999). If x_i indicates the fraction of the population knowing the word W_i, the population dynamics of such a word reads:

$$\frac{dx_i}{dt} = R_i x_i (1 - x_i) - x_i \tag{15.1}$$

with $i = 1, \ldots, n$. The first term on the right-hand side of the this equation introduces the way words are learned. The second deals with deaths of individuals at a fixed rate (here normalized to one). The way words are learned involves a nonlinear term where the interactions between those individuals knowing W_i (a fraction x_i) and those ignoring it (a fraction $1 - x_i$) are present. The parameter R_i introduces the rate at which learning takes place.

Two possible equilibrium points are allowed, obtained from $dx_i/dt = 0$. The first is $x_i^* = 0$ and the second

$$x_i^* = 1 - \frac{1}{R_i} \tag{15.2}$$

The first corresponds to the extinction of W_i (or its inability to propagate) whereas the second involves a stable population knowing W_i. The larger the value of R_i, the higher the number of individuals using the word. We can see that for a word to be maintained in the population lexicon, the following inequality must be fulfilled:

$$R_i > 1 \qquad (15.3)$$

This means that there is a threshold in the rate of word propagation to sustain a stable population. By displaying the stable population x^* against R_i (figure 15.1b), we observe a well-defined phase transition phenomenon: a sharp change occurs at $R_i^c = 1$, the critical point separating the two possible phases. The subcritical phase $R_i < 1$ will inevitably lead to loss of the word.

An important implication of this "epidemic" of word spreading is that words used below threshold will disappear. Their extinction might be slow, starting from their disappearance from the social context and surviving in dictionaries, only to eventually be removed altogether.[2]

15.3 String Models

A step further in understanding language dynamics would make use of lists of words. This is obviously an oversimplification, since language is not just a collection of objects but also a complex set of rules relating and combining them (Solé et al. 2010b). Here we ignore the fundamental role of grammar and reduce a given language to a population of users sharing the same set of words.

A fruitful toy model of language change is provided by the string approximation (Stauffer et al. 2006; Zanette 2008). In this approach, each language $\mathcal{L}_i$ is treated as a binary string, that is, $\mathcal{L}_i = (S_1^i, S_2^i, \ldots, S_L^i)$ of length L. Here $S_j^i \in \{0, 1\}$, and, as defined, a finite but very large set of potential languages exist. Specifically, a set of languages $\mathcal{L}$ is defined, namely

$$\mathcal{L} = \{\mathcal{L}_1, \mathcal{L}_2, \ldots, \mathcal{L}_M\} \qquad (15.4)$$

---

[2] For examples, see http://www.savethewords.org.

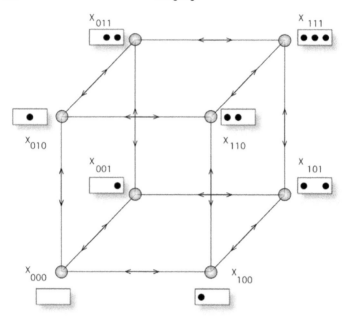

Figure 15.2. String language model. Here a given set of elements defines a language. Each (possible) language is defined by a string of ν bits (here $L = 3$) and thus 2^L possible languages are present in the hypercube. The two types of elements are indicated as filled (1) and gray (0) circles, respectively.

with $M = 2^L$. These languages can be located as the vertices of a hypercube, as shown in figure 15.2 for $L = 3$. Nodes (languages) are linked through arrows (in both directions) indicating that two connected languages differ in a single bit. This is a very small system. As L increases, a combinatorial explosion of potential strings takes place.

A given language $\mathcal{L}_i$ is shared by a population of speakers, to be indicated as x_i, and such that the total population of speakers using any language is normalized (i.e., $\sum_i x_i = 1$). A mean field model for this class of description has been proposed by

Damian Zanette, using a number of simplifications that allow
an understanding of the qualitative behavior of competing and
mutating languages (Zanette 2008). A few basic assumptions are
made in order to construct the model. First, a simple fitness
function $\phi(x)$ is defined. This function measures the likelihood
of abandoning a language. This is a decreasing function of x,
such that $\phi(0) = 1$ and $\phi(1) = 0$. Different choices are possible,
including for example $1 - x$, $1 - x^2$ or $(1 - x)^2$. On the other
hand, mutations are also included: a given language can change if
individuals modify some of their bits.

The mean field model considers the time evolution of
populations assuming no spatial interactions. If we indicate
$\mathbf{x} = (x_1, \ldots, x_M)$, the basic equations will be described in terms
of two components:

$$\frac{dx_i}{dt} = A_i(\mathbf{x}) - M_i(\mathbf{x}) \tag{15.5}$$

where both language abandonment $A_i(\mathbf{x})$ and mutation $M_i(\mathbf{x})$ are
introduced. The first term is chosen as:

$$A_i(\mathbf{x}) = \rho x_i \left(\langle \phi \rangle - \phi(x_i) \right) \tag{15.6}$$

for the population dynamics of change due to abandonment. This
is a replicator equation, where the speed of growth is defined by
the difference between average fitness $\langle \phi \rangle$, namely

$$\langle \phi \rangle = \sum_{j=1}^{N} \phi(x_j) x_j \tag{15.7}$$

and the actual fitness $\phi(x_i)$ of the i-language. Here ρ is the
recruitment rate (assumed equal to all languages). What this fitness
function introduces is a multiplicative effect: the more speakers
use a given language, the more likely it is that they will continue
to use it and that others will join the same group. Conversely, if
a given language is rare, its speakers might easily shift to some

other, more common language.

The second term includes all possible flows between "neighboring" languages. It is defined as:

$$M_i(\mathbf{x}) = \frac{\mu}{L} \left(\sum_{j=1}^{N} W_{ij} x_j - x_i \sum_{j=1}^{N} W_{ji} \right) \qquad (15.8)$$

In this sum, we introduce the transition rates W_{ij} of mutating from language $\mathcal{L}_i$ to language $\mathcal{L}_j$ and vice versa. Only single mutations are allowed, and thus $W_{ij} = 1$ if the Hamming distance $D(\mathcal{L}_i, \mathcal{L}_j)$ is exactly one. More precisely, if

$$D(\mathcal{L}_i, \mathcal{L}_j) = \sum_{k=1}^{L} |S_k^i - S_k^j| = 1 \qquad (15.9)$$

then only nearest-neighbor movements through the hypercube are allowed. In summary, $A(\mathbf{x})$ provides a description of competitive interactions whereas $M(\mathbf{x})$ gives the contribution of small changes in the string composition. The background "mutation" rate μ is weighted by the matrix coefficients W_{ij} associated with the likelihood that each specific change will occur.

This model is a general description of the bit string approximation to language dynamics. However, the general solution cannot be found and we need to analyze simpler cases. An example is provided in the next section. Although the simplifications are strong, numerical models with more relaxed assumptions seem to confirm the basic results reported below.

15.4 Symmetric Model

A solvable limit case with obvious interest to our discussion considers a population where a single language has a population x whereas all others have a small, identical size, that is, $x_i = (1-x)/(N-1)$. The main objective for defining such a *symmetric* model is to make the previous system of equations collapse into a single differential equation, which we can then analyze. In particular, we want to determine when the $x = 0$ state will be observed, meaning that no single dominant language is stable.

As defined, we have the normalization condition, now defined by:

$$\sum_{j=1}^{N} x_j = x + \sum_{j=1}^{M-1} \left(\frac{1-x}{N-1} \right) = 1 \qquad (15.10)$$

(where we choose x to be the M-th population, without loss of generality). In this case the average fitness reads:

$$\langle \phi \rangle = \phi(x)x + \sum_{j=1}^{M-1} \phi \left(\frac{1-x}{N-1} \right) \left(\frac{1-x}{N-1} \right) \qquad (15.11)$$

Using the special linear case $\phi(x) = 1 - x$, we have:

$$A(x) = \rho x(1-x) \left(x - \frac{1-x}{N-1} \right) \qquad (15.12)$$

The second term is easy to obtain: since x has (as any other language) exactly L nearest neighbors, and given the symmetry of our system, we have:

$$B(x) = \frac{\mu}{L} \left(L\frac{1-x}{N-1} - xL \right) = -\mu \left(\frac{Nx-1}{N-1} \right) \quad (15.13)$$

And the final equation for x is thus, for the large-N limit (i.e., when $N \gg 1$):

$$\frac{dx}{dt} = \rho x^2(1-x) - \mu x \qquad (15.14)$$

This equation describes an interesting scenario where growth is not logistic, as happened with our previous model of word propagation. As we can see, the first term on the right-hand side involves a quadratic component, indicating a self-reinforcing phenomenon. This type of model is typical of systems exhibiting cooperative interactions, and an important characteristic is its hyperbolic dynamics: instead of an exponential-like approximation to the equilibrium state, a very fast approach occurs.

The model has three equilibrium points: (a) the extinction state, $x^* = 0$ where the large language disappears; (b) two fixed points $x^*_\pm$ defined as:

$$x^*_\pm = \frac{1}{2} \left(1 \pm \sqrt{1 - \frac{4\mu}{\rho}} \right) \qquad (15.15)$$

As we can see, these two fixed points exist provided that $\mu < \mu_c = \rho/4$. Since three fixed points coexist in this domain of parameter space, and the trivial one ($x^* = 0$) is stable, the other two points, namely x^*_- and x^*_+, must be unstable and stable, respectively. If $\mu < \mu_c$, the upper branch x^*_+ corresponding to a monolingual solution, is stable.

In figure 15.3a we illustrate these results by means of the bifurcation diagram using $\rho = 1$ and different values of μ. In terms of the potential function we have:

$$\Phi_\mu(x) = -\rho \left(\frac{x^3}{3} + \frac{x^4}{4} \right) + \mu \frac{x^2}{2} \qquad (15.16)$$

In figure 15.3a–d three examples of this potential are shown, where we can see that the location of the equilibrium point is shifted from the monolanguage state to the diverse state as μ is tuned. The corresponding phases in the (ρ, μ) parameter space are shown in figure 15.3b.

15.5 Language Thresholds

It is interesting to see that this model and its phase transition is somewhat connected to the error threshold problem associated to the dynamics of RNA viruses. In order for a single language to maintain its dominant position, it must be efficient in recruiting and holding speakers. But it also needs to keep heterogeneity (resulting from "mutations") at a reasonably low level. If changes go

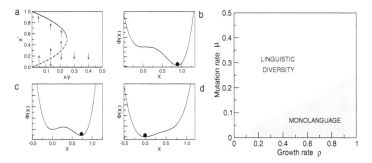

Figure 15.3. Phase transitions as bifurcations in Zanette's mean field model of language competition. In (a) we show the bifurcation diagram using μ/ρ as the control parameter. Once we cross the critical point, a sharp transition occurs from monolanguage to language diversity. This transition can be visualized using the potential function $\Phi_\mu(x)$ whose minima correspond to possible equilibrium points. Here we use $\rho = 1$ with (b) $\mu = 0.1$, (c) $\mu = 0.2$, and (d) $\mu = 0.3$.

beyond a given threshold, there is a runaway effect that eventually pushes the system into a variety of coexisting sublanguages. An error threshold is thus at work, but in this case the transition is of first order. This result would indicate that, provided a source of change is active and beyond the threshold, the emergence of multiple unintelligible tongues would be expected.

String models of this type only capture one layer of word complexity. Perhaps future models will consider ways of introducing further internal layers of organization described in terms of *superstrings*. Such superstring models should be able to introduce semantics, phonology, and other key features that are considered relevant. An example of such is provided by models of the emergence of linguistic categories (Puglisi et al. 2008).

The previous models are just a toy representation of language complexity. Human language is the outcome of both social and evolutionary pressures together with cognitive constraints

(Christiansen and Chater 2008). Brains and languages (as well as languages and societies) coevolve (Deacon 1997). In this sense, the models presented above deal with language dynamics without the explicit introduction of cognition. Although our picture can be useful in dealing with the ecology of language (Solé et al. 2010b), it fails to include the adaptive potential of communicating systems themselves. Moreover, languages cannot be encapsulated in terms of word inventories. Words interact within networks (Solé et al. 2010a) related to the correlations established among them (phonologic, syntactic, or semantic). Phase changes play a role here, too, from the emergence of word-word combinations (Nowak and Krakauer 1999) to language acquisition in children (Corominas-Murtra et al. 2009).

16

SOCIAL COLLAPSE

16.1 Rise and Fall

The reasons for the disintegration of social order in ancient civilizations have been a recurrent concern for archaeologists, historians, and economists alike (Tainter 1988; Sabloff 1997; Fagan 2004; Diamond 2005; Turchin 2006). Social collapse involves some kind of (historically) sudden, major loss of economic and social complexity. The fall of the Roman Empire, the collapse of the Mesopotamian centers, and the loss of the Mayan civilization are well-known examples of such large-scale decay (Tainter 1988; Yofee and Cowgill 1988). In all these examples, collapse implies a transition toward a simpler, less organized society that can even disintegrate and vanish. Information flows and economic transactions among different subsystems decrease or even stop, and social coherence shrinks.

The laws that drive social collapse (if such general laws exist) are not known yet, but many clues can be recovered from the analysis of the past. An example is provided by the study of the Chaco Canyon site, in the northwest corner of New Mexico (Axtell et al. 2002; Kohler et al. 2005). This was the land of the Anasazi, and is today a semiarid ecosystem of undeniable beauty, but a hard place to live. The site contains the ruins of several sets

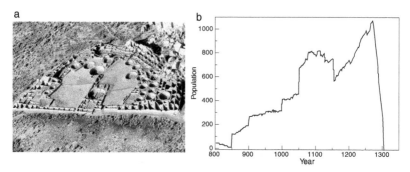

Figure 16.1. (a) An example of the large stone buildings of Chaco Canyon (picture by Bob Adams, see http://en.wikipedia.org/wiki/ PuebloBonito). It shows the basic organization of the Great House of Pueblo Bonito and involves hundreds of rooms (to be inhabited and for other purposes). In (b) the fluctuations in Anasazi populations in Long House Valley (Arizona). Population estimates are based on the number of house sites at a given time, assuming five people per house. Broad changes are observed, with a trend of increase marked by rapid shifts and a catastrophic decay between years 1250 and 1300, when the Anasazi completely abandoned the valley (redrawn from Axtell et al. 2002).

of buildings (figure 16.1a) connected to other distant locations through a network of roads spanning many miles.

Until its disappearance around 1350, the Anasazi society was marked by fluctuations in total population size, mean settlement size, and preferred habitat. It is generally agreed that the reasons for such fluctuations (as it happens with the Mayan case) depend partly on rainfall and on groundwater levels, variations in either of which would affect Anasazi maize-based agriculture. Indirect measures of both variables have been obtained for some areas in northeastern Arizona (Figure 16.1b), together with archaeological surveys of the numbers and distribution of houses for the entire span of the valley's Anasazi occupation. But how can one assess whether the proxies explain the survey results? To give just a hint of the complexities involved: readily available groundwater can make rainfall unnecessary for maize agriculture, but the dependence of maize growth on rainfall and groundwater levels varies among the valley's six habitats used for agriculture.

It might sound like a bold idea trying to compress historical dynamics of any kind into a single equation of the form $dx/dt = f(x)$. But properly formulated models and well-chosen variables have been shown to capture the key features of social and ecological dynamics (Levin 1999; Diamond 2002; Turchin 2003). The next example provides a good illustration of this point.

16.2 Slow Responses and Sunk-Cost Effects

The model presented here is based on the work described in (Janssen et al. 2003). This study provided a rationale for the presence of both resource depletion and failure to adapt in ancient societies. These models are based on the presence of a well-known deviation from rational decision making. Such a scenario is known as the *sunk-cost effect* (Arkes and Ayton 1999). The key ingredient in this model approach is an important feature of this effect: the fact that political decisions in social groups are achieved by some form of consensus. But once consensus is reached, the group will tend to commit to the decision taken, even if negative results are at work.

Such a situation can generate an important and dangerous inertia. Quoting Jansen et al.: "even when the group is faced with negative results, members may not suggest abandoning an earlier course of action, since this might break the existing unanimity."

More generally, the underlying problem here is why complex societies might fail to adapt to resource depletion. Even if there is some social perception of risk, short-term thinking often prevails when facing long-term vulnerabilities. Such undesirable behavior is often favored by a combination of incomplete understanding of the problem, together with the misleading view that all changes are reversible.

Slow social responses to ongoing problems while in their early phases can result in a high overall cost (Scheffer et al. 2003). More precisely, lags in social response to a changing world can cause late, sudden shifts. Thus societies that respond slowly to challenges might be unable to avoid the long-term consequences. Delays between recognition and regulation can result from conflicts among individuals having different perceptions, social positions, and wealth. The inertia that results from these internal conflicts can easily lead to delayed decision-making dynamics.

The consequence of the scenario described above is that there is no adaptive response to the changing resource and as a consequence the global dynamics can be described in terms of a resource-consumer system where resources are exploited with no particular forecast of future changes. As will be shown below, collapse can be a natural outcome of such a trend. The main threat here is a growing human population blindly exploiting limited resources.

16.3 Ecological Model of Social Collapse

Let us consider a simple description of the sunk-cost effect as defined by a one-dimensional model. For our analysis we take the approach used in (Janssen et al. 2002). The equation considers only resources x as the key variable and human population H is a fixed (but tunable) parameter:

$$\frac{dx}{dt} = \mu x \left(1 - \frac{x}{K}\right) - \Gamma(x)H + \delta \qquad (16.1)$$

where μ and K are the growth rate and maximum level of the resource, H is the current human population exploiting this resource, and δ indicates some resupply of x from neighboring areas.

The functional response $\Gamma(x)$ should introduce the way resources are depleted by humans, who act as consumers having a given per capita rate of consumption C. Janssen et al. use the following response:

$$\Gamma(x) = \frac{Cx}{\rho + x} \qquad (16.2)$$

where ρ is the resource level needed to reach 50 percent of the maximum consumption rate. The fixed points of this dynamical system can be obtained, as discussed in chapter 2 by looking at the intersection of two curves:

$$y_1(x) = \mu x \left(1 - \frac{x}{K}\right) + \delta \quad y_2(x) = \Gamma(x)H \qquad (16.3)$$

Those x^* such that $y_1(x^*) = y_2(x^*)$ will be the equilibrium points.

For simplicity, let us consider the case $\delta = 0$ (i.e., no external resupply). Such a simplification allows us to easily find the fixed points. The first corresponds to the (trivial) state $x^* = 0$ representing the complete depletion of resources. The stability of this fixed point can be determined from the sign of:

$$\lambda_\mu = \mu \left(1 - \frac{2x}{K}\right) - \frac{\rho CH}{(\rho + x)^2} \qquad (16.4)$$

We obtain $\lambda_\mu(0) = \mu - CH/\rho$, which will be negative if the human population density exceeds a threshold:

$$H > \bar{H} = \frac{\rho\mu}{C} \qquad (16.5)$$

The other two points are obtained from

$$\mu \left(1 - \frac{x^*}{K}\right) - \frac{CH}{\rho + x^*} = 0 \qquad (16.6)$$

which gives two possible solutions, namely

$$x_{\pm}^* = \frac{1}{2} \left[K - \rho \pm \sqrt{(K - \rho)^2 + 4K \left(\rho - \frac{CH}{\mu}\right)} \right] \qquad (16.7)$$

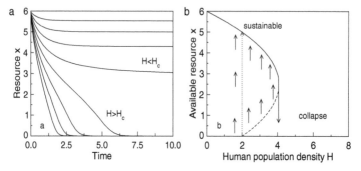

Figure 16.2. Phase transition in the sunk-cost model. In (a) we show several curves corresponding to the transient dynamics of $x(t)$ for increasing values of H (growing from top to bottom). At a given critical value H_c, the available resources collapse. The corresponding bifurcation diagram is shown in (b). Here we have used $K = 6$, $\rho = 1$, $\mu = 2$, and $C = 1$, which gives a bifurcation for $H_c = 4.08$. For $H < \bar{H} = 2$ the unstable branch does not exist and small values of x lead to a rapid flow towards high x^*.

The two fixed points $x^*_\pm$ will exist provided that the sum inside the square root is positive (otherwise it would give imaginary values). This leads to a critical value of H that defines the domain of existence of nonzero resources:

$$H_c = \frac{\mu(K + \rho)^2}{4KC} \qquad (16.8)$$

The fixed point x^*_+ is associated with a large amount of resources and thus a sustainable human population. In figure 16.2a we show several examples of the trajectories followed by our system for both $H > H_c$ and $H < H_c$. But what about the intermediate one, x^*_-? It corresponds to an unstable fixed point, separating two alternative states, as indicated in figure 16.2b with a dashed line. We can see that the only fixed point for $H < \rho\mu/C$ is actually x^*_+, thus indicating that a small amount of resource would allow a rapid recovery.

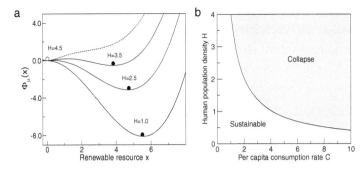

Figure 16.3. Transitions from sustainable exploitation to collapse. In (a) the potential functions associated to the sunk-cost model are shown, for three different values of the bifurcation parameter H (other parameters as in the previous figure). In (b) the two phases exhibited by the sunk-cost model. The critical line H_c is represented against the per capita consumption rate C. The gray area corresponds to the collapse of available resources.

The shifting dynamics between sustainability to collapse can be described by means of the potential $\Phi(x)$, which in our case reads:

$$\Phi(x) = -\frac{\mu}{2}x^2 + \frac{\mu}{3K}x^3 - CH\left[x + \rho\ln(\rho + x)\right] \quad (16.9)$$

This potential function is displayed in figure 16.3a for different values of human population densities H. As this parameter grows, there is a continuous change in the depth of the valley associated to the nonzero state. Once the threshold is crossed, there is a transition from a finite resource level to its collapse (figure 16.3a). The critical boundary (16.8) allows defining the two key phases for our system (figure 16.3b). These are sustainable growth and social collapse, respectively. Either increasing population size at a fixed consumption rate or introducing higher consumption at a given population level (i.e., more efficiency in resource exploitation) will eventually push the system into the collapse phase. Although we have assumed that no external income of resources is present (i.e., $\delta = 0$) the basic conclusions reached here remain the same. As the reader can check, introducing increasing levels of resource income allows the lower branch to involve a finite amount of stable resources.

16.4 Historical Dynamics

The previous model needs to be taken (as all models analyzed here) as a first approximation to reality. It is more likely that the patterns of collapse and the conditions for sustainable development are themselves nonlinear functions of different social features. The model considered here lumps together a wide range of economic and sociologic traits. It also reduces the potential diversity of available resources to a single, average resource compartment. Nevertheless, the study of the past and the possibly unavoidable presence of rapid responses to continuous changes suggest that the basic message is robust.

What might be missing here? One important component is the presence of extreme events associated with the fluctuations in external factors. This was the case when a series of severe droughts affected the Yucatan Peninsula and led to multiyear periods of very low precipitation. The Maya collapse seems correlated with several waves of drought (Gill 2000). Coupled with population growth near the limits of available resources, environmental degradation and warfare combined to doom Mayan cities (Diamond 2005). Not a single reason seems enough to fully account for this outcome. But in general, a basic pattern can be delineated involving the previous ingredients plus social inertia.

Collapse is only part of the historical dynamics of social and economic structures (Turchin 2006; Krugman 1996; Bouchaud and Cont 1998; Lansing 2005). The rules governing social dynamics have been a rich source of theoretical and statistical analysis. Statistical physics and agent-based approaches have been particularly useful in providing insight (Axelrod 1984; Epstein and Axtell 1996; Weidlich 2000; Bettencourt et al. 2007; Ball 2008). Many aspects of social organization can be measured and scrutinized by means of appropriate theoretical models. Examples include the distribution of wealth, city formation, public opinion, economic fluctuations, spatial segregation or war (Choi and Bowles 2007). Is it possible to formulate a quantitative theory of history? Some authors suggest that such dynamical theory could be a reality (Homer-Dixon 2006; Turchin 2007). In particular, an evolutionary view of cooperation and conflict among individuals suggests that the former helps building societies whereas the second (once societies become empires) favors their dissolution. If true, our view of history as a dynamical, large-scale phenomenon, would benefit from incorporating the concept of phase transitions and their associated warning signals.

REFERENCES

Adam, J. A., and Bellomo, N. 1997. *A Survey of Models for Tumor-Immune System Dynamics*. Boston, Birkhauser.

Adami, C. 2000. *Introduction to Artificial Life*. New York, Springer.

Aguirre, J., Buldú, J. M., and Manrubia, S. C. 2009. Evolutionary dynamics on networks of selectively neutral genotypes: Effects of topology and sequence stability. *Physical Review E* 80, 066112.

Albert, R., and Barabási, A. L. 2002. Statistical mechanics of complex networks. *Reviews of Modern Physics* 74, 47–97.

Alberts, B., et al. 2002. *Molecular Biology of the Cell*, 4th edition. Garland Science, New York.

Aldana, M., and Cluzel, P. 2003. A natural class of robust networks. *Proceedings of the National Academy of Sciences USA* 100, 8710–714.

Altmeyer, S., and McCaskill, J. S. 2001. Error threshold for spatially resolved evolution in the quasispecies model. *Physical Review Letters* 86, 5819–22.

Alvarez-Buylla, E., Benitez, M., Balleza-Dávila, E., Chaos, A., Espinosa-Soto, C., and Padilla-Longoria, P. 2006. Gene regulatory network models for plant development. *Current Opinion Plant Development* 10, 83–91.

Amaral, L. A. N., Scala, A., Barthandelandemy, M., and Stanley, H. E. 2000. Classes of small-world networks. *Proceedings of the National Academy of Sciences* 97, 11149–52.

Amarasekare, P. 1998. Allee effects in metapopulation dynamics. *American Naturalist* 298–302.

Anderson, A. R. A., and Quaranta, V. 2008. Integrative mathematical oncology. *Nature Reviews Cancer* 8, 227–34.

Anderson, P. W. 1972. More is different. *Science* 177, 393–96.

Anderson, R. M., and May, R. M. 1991. *Infectious Diseases of Humans. Dynamics and Control.* Oxford, Oxford University Press.

Antal, T., Krapivsky, P., and Redner, S. 2007b. Dynamics of microtubule instabilities. *Journal of Statistics Mechanics* 1, L05004.

Antal, T., Krapivsky, P., Redner, S., Mailman, M., and Chakraborty, B. 2007a. Dynamics of an idealized model of microtubule growth and catastrophe. *Physical Review E* 76, 041907.

Arkes, H. R., and Ayton, P. 1999. The sunk cost and concorde effects: Are humans less rational than lower animals. *Psychological Bulletin* 125, 591–600.

Arnold, V. 1984. *Catastrophe Theory.* Springer, Berlin.

Arthur, B. 1994. *Increasing Returns and Path Dependence in the Economy.* Ann Arbor, University of Michigan Press.

Arthur, B., Durlauf, S. N., and Lane, D. A. 1997. Introduction. In: *The Economy as an Evolving Complex System II*, W. B. Arthur, S. N. Durlauf, and D. A. Lane, eds., Santa Fe Institute Studies in the Science of Complexity, volume XXVII. Reading, MA, Addison-Wesley.

Atkins, 1995. *Physical Chemistry.* Oxford, UK, Oxford University Press.

Attneave, F. 1971. Multistability in perception. *Scientific American* 225, 62–71.

Axelrod, R. 1984. *The Complexity of Cooperation.* New York, Basic Books.

Axelrod, R., Axelrod, D., and Pienta, K. 2007. Evolution of cooperation among tumor cells. *Proceedings of the National Academy of Sciences USA* 103, 13474–79.

Axtell, R. L., Epstein, J. M., Dean, J. S., Gumerman, G. J., Swedlund, A. C., Harburger, J., Chakravarty, S., Hammond, R., Parker, J., and Parker, M. 2002. Population growth and collapse in a multiagent model of the Kayenta Anasazi in Long House Valley. *Proceedings of the National Academy of Sciences USA* 99, 7275–79.

Bak, P. 1997. *How Nature Works*. New York, Springer.

―――― 1999. *Life's Matrix*. New York, Farrar, Strauss and Giroux.

Ball, P. 2008. *Critical Mass: How One Thing Leads to Another*. New York, Farrar, Strauss and Giroux.

Barabási, A.-L., and Stanley, H. E. 1995. *Fractal Concepts in Surface Growth*. Cambridge, UK, Cambridge University Press.

Barton, N. H., Briggs, D. E. G., Eisen, J. A., Goldstein, D. B., and Patel, N. H. 2007. *Evolution*. New York, Cold Spring Harbor.

Bascompte, J., and Solé, R. V. 1995. Rethinking complexity: modelling spatiotemporal dynamics in ecology. *Trends in Ecology and Evolution* 10, 361–66.

―――― 1996. Habitat fragmentation and extinction thresholds in spatially explicit metapopulation models. *Journal of Animal Ecology* 65, 465–73.

Bascompte, J., and R. V. Solé, eds. 1998a. *Modeling Spatiotemporal Dynamics in Ecology*. Berlin, Springer-Verlag.

Beckers, R., Deneubourg, J. L., and Goss, S. 1993. Modulation of trail laying in the ant Lasius niger (Hymenoptera: Formicidae) and its role in the collective selection of a food source. *Journal of Insect Behavior* 6, 751–59

Beekman, M., Sumpter, D. J. T., and Ratnieks, F. L. W. 2001. Phase transition between disordered and ordered foraging in Pharaoh's ants. *Proceeding of the National Academy of Science USA* 98, 9703–706.

Ben-Jacob, E. 2003. Bacterial self-organization: Co-enhancement of complexification and adaptability in a dynamic environment. *Philosophical Transactions of The Royal Society of London*. A, 361, 1283–312.

Ben-Jacob, E., Becker, I., and Shapira, Y. 2004. Bacterial linguistic communication and social intelligence. *Trends in Microbiology* 12, 366–72.

Bettencourt L., Lobo, J., Helbing, D., Kuhnert, C., and West, G. B. 2007. Growth, innovation, scaling, and the pace of life in cities. *Proceeding of the National Academy of Science USA* 104, 7301–306.

Bickerton, D. 1990. *Language and Species*. Chicago, University of Chicago Press.

Binney, J. J., Dowrick, N. J., Fisher, A. J., and Newman, M. E. J. 1994. *The Theory of Critical Phenomena: An Introduction to the Renormalization Group.* Oxford, UK, Oxford University Press.

Boer, R. J., de. 1989. Extensive percolation in reasonable idiotypic networks. In: Atlan H. and Cohen I., eds., *Theories of Immune Networks.* Berlin, Springer-Verlag.

Boi, S., Couzin, I. D., Del Buono, N., Franks, N. R., and Britton, N. F. 1998. Coupled oscillators and activity waves in ant colonies. *Proceedings of the Royal Society of London B* 266, 371–78.

Bolding, K., Fulgham, M. L., and Snyder, L. 1999. *Technical report.* CSE-94-02-04.

Bollobas, B. 1985. *Random Graphs.* London, Academic Press.

Bonabeau, E. 1996. Marginally stable swarms are flexible and efficient. *Journal of Physics* 6, 309–24.

Bonabeau E., Theraulaz G., and Deneubourg J.-L. 1995. Phase diagram of a model of self-organizing hierarchies. *Physica* A 217, 373–92.

Bond, W. J., and Keeley, J. E, 2005. Fire as global "herbivore" The ecology and evolution of flammable ecosystems. *Trends in Ecology and Evolution* 20. 387–94.

Bornholdt, S. 2005. Less is more in modeling large genetic networks, *Science* 310, 449.

——— 2008. Boolean network models of cellular regulation: Prospects and limitations, *Journal of the Royal Society Interface* 5 (2008) S85–S94.

Bornholdt, S., and Schuster, H. G., eds. 2002. *Handbook of Graphs and Networks.* Berlin, Springer.

Bornholdt, S., and Wagner, F. 2002. Stability of money: Phase transitions in an Ising economy. *Physica* A 316, 453–68.

Bouchaud, J. P., and Cont, R. 2002. A Langevin approach to stock market fluctuations and crashes. *European Physical Journal* B 6, 543–50.

Bray, D. 1995. Protein molecules as computational elements in living cells. *Nature* 376, 307–12.

Bruce A., and Wallace D. 1989. Critical point phenomena: Universal physics at large length scales. In: *The New Physics*, Paul Davies, ed. Cambridge, UK, Cambridge University Press.

Brush, S. G. 1967. History of the Lenz Ising Model. *Reviews of Modern Physics* 39, 883–93.

Buchanan, M. 2007. *The Social Atom*. New York, Bloomsbury Press.

Buhl, J., Sumpter, D. J. T., Couzin, I. D., Hale, J. J., Despland, E., Miller, E. R., and Simpson, S. J. 2006. From disorder to order in marching locusts. *Science* 312, 1402–06.

Bull, J. J., Meyers, L. A., and Lachmann, M. 2005. Quasispecies made simple. *PLOS Computational Biology* 1, e61.

Bussemaker, H. J., Deutsch, A., and Geigant, E. 1997. Mean-field analysis of a dynamical phase transition in a cellular automaton model. *Physical Review Letters* 78, 5018–21.

Byrne, H. M., Alarcon, T., Owen, M. R., Webb, S. D., and Maini, T. K. 2006. Modelling aspects of cancer dynamics: A review. *Phil Trans R Soc* A 364, 1563–78.

Cahill, D. P., Kinzler, K., Vogelstein, B., and Lengauer, C. 1999. Trends in Genetic instability on Darwinion sele 15, M57–M60.

Cairns, J. 1975. Mutation, selection and the natural history of cancer. *Nature* 255, 197–200.

Camazine, S., Deneubourg, J., Franks, N. R., Sneyd, J., Theraulaz, G., and Bonabeau, E. 2002. *Self-Organization in Biological Systems*. Princeton, Princeton University Press.

Carroll, S. B. 2001. Chance and necessity: The evolution of morphological complexity and diversity. *Nature* 409, 1102–09.

Case, T. J. 2000. *An Illustrated Guide to Theoretical Ecology*. New York, Oxford University Press.

Chaikin, P. M., and Lubensky, T. C. 1995. *Principles of Condensed Matter Physics*. Cambridge UK, Cambridge University Press.

Chaos, A., Aldana, M., Espinosa-Soto, C., Garcia Ponce, B., Garay-Arroyo, A., and Alvarez-Buylla, E. R. 2006. From genes to flower patterns and evolution: Dynamic models of gene regulatory networks. *Plant Growth Regulation* 25, 278–89.

Chisholm, R., and Filotas, E. 2009. Critical slowing down as an indicator of transitions in two-species models. *Journal of Theoretical Biology* 257, 142–49.

Choi, J-K., and Bowles, S. 2007. The coevolution of parochial altruism and war. *Science* 318, 636–40.

Christiansen, M. H., and Chater, N. 2008. Lauguage as shaped by the brain. *Behavioral and Brain Sciences* 31, 489–558.

Christensen, K., and Moloney, N. R. 2005. *Complexity and Criticality.* London, Imperial College Press.

Cochrane M. A., 2003. Fire science for rainforest. *Nature* 421, 913–19.

Codoner, F. M., Daros, J. A., Sole, R. V., and Elena, S. F. 2006. The fittest versus the flattest: Experimental confirmation of the quasispecies effect with subviral pathogens. *PLoS Pathog* 2, e136.

Cole, B. J. 1991a. Short-term activity cycles in ants: Generation of periodicity by worker interaction. *The American Naturalist* 137, 244–59.

——— 1991b. Is animal behaviour chaotic? Evidence from the activity of ants. *Proceedings of the Royal Society, London.* Series B. 244, 253–59.

——— 1991c. Short-term activity cycles in ants: A phase response curve and phase resetting in worker activity. *The Journal of Insect Behavior* 4, 129–37.

Colizza, V., Barrat, A., Barthelemy, M., Valleron, A.-J., and Vespignani, A. 2007. Modeling the worldwide spread of pandemic influenza: Baseline case and containment interventions. *PLoS Medicine* 4(1), e13.

Corominas-Murtra, B., Valverde, S., and Solé, R. V. 2009. The ontogeny of scale-free syntax networks: Phase transitions in early language acquisition. *Advances in Complex Systems* 12, 371–92.

Courchamp, F., Angulo, E., Rivalain, P., Hall, R., Signoret, L., Bull, L., and Meinard, Y. 2006. Rarity value and species extinction: The authropogenic Allee Effect. *PloS Biology* 4, e415.

Coveney, P., and Highfield, J. 1989. *The Arrow of Time.* New York, Flamingo.

——— 1995. *Frontiers of Complexity: The Search for Order in a Chaotic World.* New York, Faber Inc.

Cowan, G. A., Pines, D., and Meltzer, D. 1994. *Complexity: Metaphors, Models, and reality.* Reading, MA, Addison-Wesley.

Cristini V., Frieboes, H. B., Gatenby, R., Caserta, S., Ferrari, M., and Sinek, J. 2005. Morphologic instability and cancer invasion. *Clinical Cancer Research* 11, 6772–79.

Crystal, D. 2000. *Language Death*. Cambridge, UK, Cambridge University Press.

Cui, Z., Willingham, M. C., Hicks, A. M., Alexander-Miller, M. A., Howard, T. D., Hawkins, G. A., Miller, M. S., Weir, H. M., Du, W., and DeLong, C. J. 2003. Spontaneous regression of advanced cancer: Identification of a unique genetically determined, age-dependent trait in mice. *Proceedings of the National Academy of Sciences USA* 100, 6682–87.

Dakos, V., Scheffer, M., van Nes, E. H., Brovkin, V., Petoukhov, V., and Held, H. 2008. Slowing down as an early warning signal for abrupt climate change. *Proceedings of the National Academy of Sciences USA* 105, 14308–12.

Davidich, M. I., and Bornholdt, S. 2008. The transition from differential equations to Boolean networks: A case study in simplifying a regulatory network model. *Journal Theorotic Biology* 255, 269–77.

Deacon, T. 1997. *The Symbolic Species: The Co-Evolution of Language and the Brain*. New York, Nortons.

Deakin, M. A. B., 1980. Applied catastrophe theory in the social and biological sciences. *Bulletin of Mathematical Biology* 42, 647–79.

Delgado, J., and Solé, R. V. 1997. Noise–induced transitions in fluid neural networks. *Physics Letters* A 229, 183–89.

——— 2000. Self-synchronization and task fulfilment in ant colonies. *Journal of Theoretical Biology* 205, 433–41.

Deneubourg, J. L., and Goss, S. 1989a. Collective patterns and decision-making. *Ethology, Ecology and Evolution* 1, 295–311.

Deneubourg, J. L., Goss, S., Franks, N., and Pasteels, J. M. 1989b. The blind leading the blind: Modeling chemically mediated army ant raid patterns. *Journal of Insect Behavior* 2, 719–25.

Deneubourg, J. L., Pasteels, J. M., and Verhaeghe, J. C. 1983. Probabilistic behaviour in ants: A strategy of errors? *Journal of Theoretical Biology* 105, 259–71.

Dennis, B. 1989. Allee effects: Population growth, critical density, and the chance of extinction. *Natural Resource Modeling* 3, 481–538.

Derrida, B., and Pomeau, Y. 1986. Random networks of automata: A simple annealed approximation. *Europhysics Letters* 1, 45–49.

Derrida, B., and Stauffer, D. 1986. Phase transitions in two-dimensional Kauffman cellular automata. *Europhysics Letters* 2, 739–45.

Detrain, C., and Deneubourg, J. L. 2006. Self-organised structures in a superorganism: Do ants behave like molecules? *Physics of Life Reviews* 3, 162–87.

Dezso, Z., and Barabási, A.-L. 2002. Halting viruses in scale-free networks. *Physical Review E* 65, 055103.

Diamond, J. 1997. *Guns, Germs and Steel: The Fates of Human Societies.* New York, Random House.

——— 2005. *Collapse: How Societies Choose to Fail or Succeed.* New York, Viking Books.

Ditzinger T., and Haken, H. 1989. Oscillations in the perception of ambiguous patterns. *Biological Cybernetics* 61, 279–87.

Dogorovtsev, S. N., and Mendes, J. F. F. 2003. *Evolution of Networks.* Oxford, Oxford University Press.

Dormingo, E. (editor) 2006. *Quasispecies: Concepts and Implications for Virology.* Heidelberg, Springer.

Duke, T. A. J, and Bray, D. 1999. Heighted sensitivity of a lattice of membrane receptors. *Proceedings of the National Academy of Sciences USA* 96, 10104–108.

Dyson, F. 1999. *Origins of Life.* New York, Oxford University Press.

Eigen, M. 1971. Self-organization of matter and evolution of biological macromolecules. *Naturwissenschaften* 58, 465–523.

Eigen, M., and Schuster, P. 1999. *The Hypercycle.* Berlin, Springer.

Eiha, N., et al. 2002. The mode transition of the bacterial colony. *Physica* A 313, 609–24.

Epstein, J. M., and Axtell, R. 1996. *Growing Artificial Societies: Social Science from the Bottom up.* Cambridge, MA, MIT Press.

Evans, D., and Wennerström, H. 1999. *The Colloidal Domain.* New York, Wiley.

Fagan, B. M. 2004. *The Long Summer: How Climate Changed Civilization.* New York, Basic Books.

Fermi, E. 1953. *Thermodynamics.* London, Dover.

Ferreira, C. P., and Fontanari, J. F. 2002. Nonequilibrium phase transitions in a model for the origin of life. *Physical Review E* 65, 021902.

Ferrer, R., and Solé, R. V. 2003. Least effort and the origins of scaling in human language. *Proceedings of the National Academy of Sciences USA* 100, 788–91.

Foley, J. A., Coe, M. T., Scheffer, M., and Wang, G. 2003. Regime shifts in the Sahara and Sahel. Interactions between ecological and climate systems in northern Africa. *Ecosystems* 6, 524–39.

Forgacs, G. 1995. On the possible role of cytoskeletal filamentous networks in intracellular signaling: An approach based on percolation. *Journal of Cell Science* 108.

Forgacs, G., and Newman, S. A. 2007. *Biological Physics of the Developing Embryo*. Cambridge UK, Cambridge University Press.

Forgacs, G., Newman, S. A., Obukhov, S. P., and Birk, D. E. 1991. Phase transition and morphogenesis in a model biological system. *Physical Review Letters* 67, 2399–402.

Fowler, M. S., and Ruxton G. D. 2002. Population dynamic consequences of Allee effects. *Journal of Theoretical Biology* 215, 39–46.

Franks, N. R., and Bryant, S. 1987. Rhythmical patterns of activity within the nests of ants. In: J. Eder and A. Rembold, Editors, *Chemistry and Biology of Social Insects*. pp. 122–23. Berlin, Springer.

Fuchs A., Kelso J. A. S., and Haken, H. 1992. Phase transitions in the human brain: Spatial mode dynamics. *International Journal of Bifurcation and Chaos* 2, 917–39.

Fuks, H., and Lawniczak, A. T. 1999. Performance of data networks with random links. *Mathematics and Computers in Simulation* 51, 101–17.

Fuller, B. 1961. Tensegrity. *Portfolio Artnews Annual* 4, 112–27.

Garay, R. P., and Lefever, R. 1977. Kinetic apprach to the immunology of cancer: Stationary state properties of effector-target cell reactions. *Journal of Theoretical Biology* 73, 417–38.

Garnier, S., Gautrais, J., and Theraulaz, G. 2007. The biological principles of swarm intelligence. *Swarm Intelligence* 1, 3–31.

Gatenby, R. A. 1996. Application of competition theory to tumour growth. *European Journal of Cancer* 32, 722–26.

Gatenby, R. A., and Frieden, R. 2002. Application of information theory and extreme physical information to carcinogenesis. *Cancer Research* 62, 3675–82.

Gautrais, J., Jost, C., and Theraulaz, G. 2008. Key behavioural factors in a self-organised fish school model. *Annales Zoologici Fennici* 45, 415–28.

Gell-Mann, M. 1994. *The Quark and the Jaguar*. New York, Freeman.

Gennes, P. de. 1976. On a relation between percolation and the elasticity of gels. *Journal de Physique* 37, L1–L2.

Gill, R. B. 2000. *The Great Maya Droughts: Water, Life, and Death*. Albuquerque, University of New Mexico Press.

Gilmore, 1981. *Catastrophe Theory*. London, Dover.

Glass, L., and Hill, C. 1998. Ordered and disordered dynamics in random networks. *Europhysics Letters* 41, 599–604.

Goldenfeld, N. 1992. *Lectures On Phase Transitions and the Renormalization Group*. New York, Westview Press.

Gonzalez-Garcia I., Sole, R. V., and Costa, J. 2002. Metapopulation dynamics and spatial heterogeneity in cancer. *Proceedings of the National Academy of Sciences USA* 99, 13085–89.

Goodwin, B. C. 1994. *How the Leopard Changed Its Spots: The Evolution of Complexity*. New York, Charles Scribner and Sons.

Gordon, D. 1999. *Ants at Work*. New York, Free Press.

Goss, S., and Deneubourg, J. L. 1988. Autocatalysis as a source of synchronised rhythmical activity in social insects. *Insectes Sociaux* 35, 310–15.

Gould, S. J. 1989. *Wonderful Life*. London, Penguin.

Grenfell, B. T., Bjornstad, O. N., and Kappey, J. 2001. Travelling waves and spatial hierarchies in measles epidemics. *Nature* 414, 716–23.

Guth, A. 1999. *The Inflationary Universe*. Reading MA, Perseus.

Haken, H. 1977. Berlin, Springer.

———— 1996. *Principles of Brain Functioning*. Berlin, Springer.

———— 2002. *Brain Dynamics*. Berlin, Springer.

———— 2006. Synergetics of brain function. *International Journal of Psychophysiology* 60, 110–24.

Haldane, A. G., and May, R. M. 2011. Systemic risk in banking ecosystems. *Nature* 469, 351–55.

Hanahan, D., and Weinberg, R. A. 2000. The hallmarks of cancer. *Cell* 100, 57–70.

Hanel, R., Kauffman, S. A., and Thurner, S. 2005. Phase transition in random catalytic networks. *Physical Review E* 72, 036117.

Hanski, I. 1999. *Metapopulation Ecology*. Oxford, Oxford University Press.

Harold, F. M. 2001. *The Way of the Cell: Molecules, Organisms and the Order of Life*. New York, Oxford University Press.

Harris, T. E. 1963. *The Theory of Branching Processes*. Berlin, Springer.

Hauser, M. D., Chomsky, N., and Fitch, W. T. 2002. The faculty of language: What is it, who has it and how did it evolve? *Science* 298, 1569–79.

Hertz, J., Krogh, A., and Palmer, R. G. 1991. *Introduction to the Theory of Neural Computation*. New York, Westview Press.

Hill, M. F., and Caswell, H. 1999. Habitat fragmentation and extinction thresholds on fractal landscapes. *Ecology Letters* 2, 121–27.

Hillis, D. 1998. *The Pattern on the Stone: The Simple Ideas that Make Computers Work*. New York, Basic Books.

Hinrichsen, H. 2000. Non-equilibrium critical phenomena and phase transitions into absorbing states. *Advances in Physics* 49, 815–958.

Hofbauer, J., and Sigmund, K. 1991. *The Theory of Evolution and Dynamical Systems*. Cambridge, Cambridge University Press.

Hofstadter, D. 1980. *Gödel, Escher, Bach*. New York, Basic Books.

Hogeweg, P. 2002. Computing an organism: On the interface between informatic and dynamic processes. *Biosystems* 64, 97–109.

Holland, J. 1998. *Emergence: From Chaos to Order*. New York, Basic Books.

Holldobler, B., and Wilson, E. O. 2009. *The Superorganism: The Beauty, Elegance and Strangeness of Insect Societies*. New York, Norton.

Homer-Dixon, T. 2006. *The Upside of Down: Catastrophe, Creativity, and the Renewal of Civilization*. Washington DC, Island Press.

Hopfield, J. J. 1994. Physics, computation and why biology looks so different. *Journal of Theoretical Biololgy* 171, 53–60.

Howard, J., and Hyman, A. A. 2003. Dynamics and mechanics of the microtubule plus end. *Nature* 422, 753–58.

Howe, C. J., Barbrook, A. C., Spencer, M., Robinson, P., Bordalejo, B., and Mooney, L. R. 2001. Manuscript evolution. *Trends in Genetics* 17, 147–52.

Huang, S. 2004. Back to the biology in systems biology: What can we learn from biomolecular networks? *Briefings in Functional Genomics and Proteomics* 2, 279–97.

Huang, S., Eichler, G., Bar-Yam, Y., and Ingber, D. 2005. Cell fate as a high-dimensional attractor of a complex gene regulatory network. *Physical Review Letters* 94, 128701.

Huang, S., Guo, Y. P., May, G., and Enver, T. 2007. Bifurcation dynamics of cell fate decision in bipotent progenitor cells. *Developmental Biology* 305, 695–713.

Huberman, B., and Hogg, T. 1987. Phase transitions in artificial intelligence systems. *Artificial Intelligence* 33, 155–71.

Huberman, B., and Lukose, B. 1997. Social dilemmas and Internet congestion. *Science* 277, 535–37.

Ingber, D. E. 1998. The architecture of life. *Scientific American* 278, 48–57.

———— 2003. Tensegrity I: Cell structure and hierarchical systems biology. *Journal of Cell Science* 116, 1157–73.

Isaacs, F. J., Hasty, J., Cantor, C. R., and Collins, J. J. 2003. Prediction and measurement of an autoregulatory genetic module. *Proceedings of the National Academy of Sciences* 100, 7714–19.

Istrail, S. 2000. *Statistical Mechanics, Three-Dimensionality and NP-Completeness: I. Universality of Intractability of the Partition Functions of the Ising Model across Non-Planar Lattices.* Proceedings of the 32nd ACM Symposium STOC00, ACM Press, pp. 87–96, Portland, OR.

Jackson, 1990a. *Perspectives of Nonlinear Dynamics.* Vol. I. Cambridge, UK, Cambridge University Press.

———— 1990b. *Perspectives of Nonlinear Dynamics.* Vol. II. Cambridge, UK, Cambridge University Press

Jain, S., and Krishna, S. 2001. Crashes, recoveries, and "core-shifts" in a model of evolving networks. *Proceedings of the National Academy of Sciences USA* 98, 543–47.

Jain, K., and Jrug, J. 2007. Adaptation in simple and complex fitness landscapes. In: *Structural Approaches to Sequence Evolution*, Ugo Bastolla, ed. pp. 299–339. Berlin, Springer.

Janeway, C. A., Travers, P., Walport, M., and Shlomchik, M. J. 2001. *Immunobiology.* New York, Garland Science.

Janssen, M. A., Kohler, T. A., and Scheffer, M. 2003. Sunk-cost effects and vulnerability to collapse in ancient societies. *Current Anthropology* 44, 722–28.

Jirsa V. K., Friedrich R., Haken H., and Kelso J. A. 1994. A theoretical model of phase transitions in the human brain. *Biological Cybernetics* 71, 27–35.

Kacperski, K., and Hoyst, J. A. 1996. Phase transitions and hysteresis in a cellular automata-based model of opinion formation. *Journal of Statistical Physics* 84, 169–89.

Kardar, M. 2007 *Statistical Physics of Fields*. Cambridge, UK, Cambridge University Press.

Karp, G. 1999. *Cell and Molecular Biology*, 2nd edition. New York, John Wiley.

Katori, M., Kizaki, S., Terui, Y., and Kubo, T. 1998. Forest dynamics with canopy gap expansion and stochastic Ising model. *Fractals* 6, 81–86.

Kauffman, S. A. 1962. Metabolic stability and epigenesis in randomly connected nets. *Journal of Theoretical Biology* 22, 437–67.

———— 1993. *The Origins of Order*. Oxford UK, Oxford University Press.

Kéfi, S. 2009. *Reading the signs. Spatial vegetation patterns, arid ecosystems and desertification*. Ph.D. thesis. Utrecht University.

Kéfi, S., Rietkerk, M., Alados, C. L., Pueyo, Y., Elaich, A., Papanastasis, V., and de Ruiter, P. C. 2007. Spatial vegetation patterns and imminent desertification in Mediterranean arid ecosystems. *Nature* 449, 213–17.

Kelso, J. A. S. 1995. *Dynamic Patterns: The Self-Organization of Brain and Behavior*. Cambridge, MA, MIT Press.

Kitano, H. 2004. Cancer as a robust system: Implications for anticancer therapy. *Nature Reviews Cancer* 4, 227–35.

Kiyono, K., Struzik, Z. R., Aoyagi, N., Togo, F., and Yamamoto, Y. 2005. Phase transition in a healthy human heart rate. *Physical Review Letters* 95, 058101.

Kiyono, K., Struzik, Z. R., and Yamamoto, Y. 2006 Criticality and phase transition in stock-price fluctuations. *Physical Review Letters* 96, 068701.

Klausmeier, C. A. 1999. Regular and irregular patterns in semiarid vegetation. *Science* 284, 1826–28.

Kleinschmidt, A., Büchel, C., Zeki, S., and Frackowiak, S. J., 1998. Human Brain Activity during Spontaneously Reversing Perception of Ambiguous Figures. *Proceedings of the Royal Society of London* B. 265, 2427–33.

Kohler, T. A., Gumerman, G. J., and Reynolds, R. G. 2005. Simulating ancient societies. *Scientific American* 293, 76–82.

Krugman, P. 1996. *The Self-Organizing Economy*. Cambridge, MA, Blackwell.

Kürten, K. 1988. Critical phenomena in model neural networks. *Physics Letters A* 129, 156–60.

Kuznetsov, Yu. 1995. *Elements of Applied Bifurcation Theory*. New York, Springer-Verlag.

Landau, K., and Binder, D. W. 2000. *Monte Carlo Simulation in Statistical Physics*. Berlin, Springer.

Landauer, R. 1961. Irreversibility and heat generation in the computing process. *IBM Journal of Research and Development* 5, 183–91.

Langton, C. G. 1990. Computation at the edge of chaos: Phase transitions and emergent computation. *Physica D* 42, 12–37.

Lansing, J. G. 2002. Artificial societies and the social sciences. *Artificial Life*, 8, 279–92.

Lenton, T. M., Held, H., Kriegler, E., Hall, J. W., Lucht, W., Rahmstorf, S., and Schellnhuber, H. J. 2008. Tipping Elements in the Earth's Climate System. *Proceedings of the National Academy of Sciences USA* 105, 1786–93.

Leung, H., Kothari, R., and Minai, A. 2003. Phase transition in a swarm algorithm for self-organized construction. *Physical Review E* 68, 46111.

Levin, S. A. 1998. Ecosystems and the biosphere as complex adaptive systems. *Ecosystems* 1, 431–36.

——— 1999. *Fragile Dominion*. New York, Perseus Books.

Levins, R. 1969. Some demographic and genetic consequences of environmental heterogeneity for biological control. *Bulletin of the Entomological Society of America* 15, 237–40.

Lewenstein, M., Nowak, B., and Latane, B. 1992. Statistical mechanics of social impact. *Physical Review A* 45, 703–16.

Liebhold, A. M., and Bascompte, J. 2003. The allee effect, stochatic dynamics and the eradication of alien species. *Ecology Letters* 6, 133–40.

Liljeros, F., Edling, C. R., Amaral, L.A.N., Stanley, H. E., and Aberg, Y. 2001. The web of human sexual contacts. *Nature* 411, 907–908.

Linde, A. 1994. The self-reproducing inflationary universe. *Scientific American* 271, 48–53.

Livio, M. 2005. *The Equation that Couldn't Be solved*. New York, Simon and Schuster.

Lloyd, A. L., and May, R. M. 2001. How viruses spread among computers and people. *Science* 292, 1316–17.

Loeb, L. A., Loeb, K. R., and Anderson, J. P. 2003. Multiple mutations and cancer. *Proceedings of the National Academy of Sciences USA* 100, 776–81.

Loengarov, A., and Tereshko, V. 2008. Phase transitions and bistability in honeybee foraging dynamics. *Artificial life* 14, 111–20.

Luque, B., and Solé, R. V. 1997. Phase Transitions in random networks: Simple analytic determination of critical points. *Physical Review E* 55, 257.

Macia, J., Widder, S., and Solé, R.V. 2009. Why are cellular switches Boolean? General conditions for multistable genetic circuits. *Journal of Theoretical Biology* 261, 126–35.

Maley, C., Odze, R. D., Reid, B. J., Rabinovitch, P. S., et al. 2006. Genetic clonal diversity predicts progression to esophageal adenocarcinoma. *Nature Genetics* 38, 468–73.

Marro, J., and Dickman, R. 1999. *Nonequilibrium Phase Transitions in Lattice Models*. Cambridge, Cambridge University Press.

Mas, A., Lopez-Galindez, C., Cacho, I., Gomez, J., and Martinez, M. A. 2010. Unfinished stories on viral quasispecies and Darwinian views of evolution. *Journal of Molecular Biology* 397, 865–77.

Matsushita, M., Wakita, J., Itoh, H., Rafols, I., Matsuyama, T., Sakaguchi, H., and Mimura, M. 1998. Interface growth and pattern formation in bacterial colonies. *Physica* A 249, 517–24.

May, R. M. 1972. Will a large complex system be stable? *Nature* 238, 413–14.

———— 1973. *Stability and Complexity in Model Ecosystems*. Princeton, Princeton University Press.

Maynard Smith, J., and Szathmáry, E. 1999. *The Major Transitions in Evolution*. Oxford, UK, Oxford University Press.

McNeil, K. J., and Walls, D. F. 1974. Chemical instabilities as critical phenomena. *Physical Review* 17, 1513–28.

———— 1990. Exploring Complexity. New York, Freeman.

McWorter, J. 2003. *The Power of Babel : A Natural History of Language*. New York, Times Books.

Menocal, P., B., de Ortiz, J., Guilderson, T., Adkins, J., Sarnthein, M., Baker, L., and Yarusinski, M. 2000. Abrupt onset and termination of the African Humid Period: Rapid climate response to gradual insolation forcing. *The Quaterly Science Review* 19, 347–61.

Merlo, J., Pepper, B., Reid, B. J., and C. Maley. 2006. Cancer as an evolutionary and ecological process. *Nature Review Cancer* 6, 924–35.

Mikhailov A. 1994. *Foundations of Synergetics, Volume 1: Distributed Active Systems*. Berlin, Springer.

Mikhailov A., and Calenbuhr V., 2002. *From Cells to Societies*. Berlin, Springer.

Miller, S. L. 1953. Production of amino acids under possible Primitive earth conditions. *Science* 117, 528.

Millonas, M. 1993. Swarms, phase transitions, and collective intelligence. In: *Artificial Life III*. C. Langton, ed. Reading, MA, Addison-Wesley.

Milne, B. T. 1998. Motivation and benefits of complex systems approaches in ecology. *Ecosystems* 1, 449–56.

Miramontes, O. 1995. Order-disorder transitions in the behavior of ant societies. *Complexity* 1, 56–60.

Mitchell, M. 2009. *Complexity: A Guided Tour*. New York, Oxford University Press.

Monasson, R., Zecchina, R., Kirkpatrick, S., Selman, B., and Troyansky, L. 1999. Determining computational complexity from characteristic "phase transitions". *Nature* 400, 133–37.

Montroll, E. W. 1981. On the dynamics of the Ising model of cooperative phenomena. *Proceedings of the National Academy of Sciences USA* 78, 36–40.

Moore, C., and Mertens, S. 2009. *The Nature of Computation.* New York, Oxford University Press.

Morowitz, H. J. 2002. *The Emergence of Everything. How the World Became Complex.* Oxford, UK, Oxford University Press.

Mouritsen, O. 2005. *Life as a Matter of Fat: The Emerging Science of Lipidomics.* Berlin, Springer.

Murray, J. D. 1989. *Mathematical Biology.* Heidelberg, Springer.

Nagel, K., and Rasmussen, S. 1994. Traffic at the edge of chaos. In: R. A. Brooks and P. Maes, eds. *Artificial Life IV*, pp. 222–35. Cambridge, MA, MIT Press.

Nagel, K., and Schreckenberger, M. J. 1992. A cellular automaton model for freeway traffic. *Journal de Physique* I. 2, 2221–9.

Newman, M. E. J. 2003a. Random graphs as models of networks. In: *Handbook of Graphs and Networks*, S. Bornholdt and H. G. Schuster, eds., Berlin, Wiley-VCH.

———— 2003b. The structure and function of complex networks. *SIAM Review* 45, 167–256.

Nicolis, G. 1995. *Introduction to Nonlinear Science.* New York, Cambridge University Press.

Nicolis, G., Malek-Mansour, M., Kitahara, K., and van Nypelseer, A. 1976. Fluctuations and the onset of instabilities in nonequilibrium systems. *Physics Letters A* 48, 217–18.

Nicolis, G., and Prigogine, I. 1977. *Self-Organization in Nonequilibrium Systems.* New-York, Wiley.

———— 1990. *Exploring Complexity.* New York, Freeman.

Nimwegen, E. van, Crutchfield, J. P., and Huynen, M. 1999. Neutral evolution of mutational robustness. *Proceedings of the National Academy of Sciences USA* 96, 9716–20.

Nimwegen, E. van, and Crutchfield, J. P. 2000. Metastable evolutionary dynamics: Crossing fitness barriers or escaping via neutral paths? *Bulletin of Mathematical Biology* 62, 799–848.

Nitzan, A. 1974. Chemical instabilities as critical phenomena. *Physical Review* 17, 1513–28.

Nowak, M., and Krakauer, D. 1999. The evolution of language. *Proceedings of the National Academy of Sciences USA* 96, 8028–33.

Ohira, T., and Sawatari, R. 1998. Phase transition in a computer network traffic model. *Physical Review E* 58, 193–95.

Onsager, L. 1944. Crystal statistics I. A two-dimensional model with an order-disorder transition. *Physical Review* 65, 117–49.

Oró, J. 1961a. Comets and the formation of biochemical components on the primitive earth. *Nature* 190, 389–90.

——— 1961b. Mechanism of synthesis of adenine from hydrogen cyanide under possible primitive earth conditions. *Nature* 191, 1193–94.

Ortoleva, P., and Yip, S. 1976. Computer molecular dynamics studies of a chemical instability. *Journal of Chemical Physics* 65, 2045–51.

Oster, G. F., and Wilson, E. O. 1978. *Caste and Ecology in the Social Insects*. Princeton, Princeton University Press.

Parisi, G. 1988. *Statistical Field Theory*. Reading, MA, Perseus Books.

Pascual, M. 2005. Computational ecology: From the complex to the simple. *PLOS Computational Biology* 1, e18.

Pastor-Satorras, R., and Solé, R. V. 2001. Field theory for a reaction-diffusion model of quasispecies dynamics. *Physical Review* E64, 051909.

Perelson A. S. 1989. Immune network theory. *Immunological Reviews* 110, 5–36.

Perelson, A. S., and Weisbuch, G. 1997 Immunology for physicists. *Reviews of Modern Physics* 69, 1219–67.

Pimm, S. L. 1991. *The Balance of Nature? Ecological Issues in the Conservation or Species and Communities*. Chicago, University of Chicago Press.

Plerou, V., Paramesvaran, G., and Stanley, H. E. 2003. Two-phase behavior of financial markets. *Nature* 421, 130.

Pollack, G. H. 2001. *Cells, Gels and the Engines of Life*. Seattle, Ebner and Sons.

Pollack, G. H., and Chin, W.-C. 2008. *Phase Transitions in Cell Biology*. Berlin, Springer.

Pollard, T. D. 2003. The cytoskeleton, cellular motility and the reductionist agenda. *Nature* 422, 741–45.

Poston, T., and Stewart, I. 1978. *Catastrophe Theory and Its Applications.* New York, Dover.

Pueyo, S. 2007. Self-organised criticality and the response of wildland fires to climate change. *Climatic Change* 82, 131–61.

Pueyo, S., et al. 2010. Testing for criticality in ecosystem dynamics: The case of Amazonian rainforest and savanna fire. *Ecology Letters* 13, 793–802.

Puglisi, A., Baronchelli, A., and Loreto, V. 2008. Cultural route to the emergence of linguistic categories. *Proceedings of the National Academy of Sciences USA* 105, 7936–40.

Ricardo A., and Szostak J. W. 2009. Origin of life on earth. *Scientific American* 301, 54–61.

Rietkerk, M., Dekker, S. C., De Ruiter, P., and Van de Koppel, J. 2004. Self-organized patchiness and catastrophic shifts in ecosystems. *Science* 305, 1926–29.

Sabloff, J. A. 1997. *New Archeology and the Ancient Maya.* New York, Freeman and Co.

Sahimi, M. 1994. *Applications of Percolation Theory.* London, Taylor and Francis.

Sanjuán, R., Cuevas, J. M., Furió, V., Holmes, E. C., and Moya, A. 2007. Selection for robustness in mutagenized RNA viruses. *PLOS Genetics*, 15, e93.

Sardanyés, J., Elena, S. F., and Solé, R. V. 2008. Simple quasispecies models for the survival-of-the-flattest effect: The role of space. *Journal of Theoretics Biology* 250, 560–68.

Scanlon, T., Caylor, K., Levin, S., and Rodriguez-Iturbe, I. 2007. Positive feedbacks promote power-law clustering of Kalahari vegetation. *Nature* 449, 209–94.

Scheffer, M., and Carpenter, S. R. 2003. Catastrophic regime shifts in ecosystems: Linking theory to observation. *Trends in Ecology and Evolution* 18, 648–56.

Scheffer, M., Carpenter, S., Foley, J. A., Folke, C., and Walker, B. 2001. Catastrophic shifts in ecosystems. *Nature* 413, 591–96.

Schlicht R., and Iwasa, Y. 2004. Forest gap dynamics and the Ising model. *Journal of Theoretical Biology* 230, 65–75.

Schulman L. S., and Seiden, P. E., 1986. Percolation and galaxies. *Science* 233, 425–31.

——— 1981. Percolation analysis of stochastic models of galactic evolution. *Journal Statistical Physics* 27, 83–118.

Schuster, P. 1994. How do RNA molecules and viruses explore their worlds? In: G. A. Cowan, D. Pines, and D. Meltzer, eds. *Complexity: Metaphors, Models and Reality*, Addison-Wesley, Reading, MA, pp. 383–418.

——— 1996. How complexity arises in evolution. *Complexity* 2, 22–30.

Schuster P., Fontana, W., Stadler, P., and Hofacker, I. 1994. From sequences to shapes and back: A case study in RNA secondary structures. *Proceedings of the Royal Society of London* B 255, 279–84.

Schweitzer F., ed. 2002. *Modeling Complexity in Economic and Social Systems*. Singapore, World Scientific.

——— 2003. *Brownian Agents and Active Particles*. Berlin, Springer.

Segel, L. A. 1998. Multiple attractors in immunology: Theory and experiment. *Biophysical Chemistry* 72, 223–30.

Shen, Z.-W. 1997. Exploring the dynamic aspect of sound change. *Journal of Chinese Linguistics*. Monograph Series Number 11.

Shneb, N. M., Sarah, P., Lavee, H., and Solomon, S. 2003. Reactive glass and vegetation patterns. *Physical Review Letters* 90, 0381011.

Sieczka, P., and Holyst, J. 2008. A threshold model of financial markets. *Acta Physica Polonica* A 114, 525.

Slack, J. M. W. 2003. C. H. Waddington—the last Renaissance biologist. *Nature Reviews Genetics* 3, 889–95.

Solé, R. V. 2003. Phase transitions in unstable cancer cell populations. *European Physics Journal* B35, 117–24.

——— 2007. Scaling laws in the drier. *Nature* 449, 151–53.

Solé, R. V., and Bascompte, J. 2006. *Self-Organization in Complex Ecosystems*. Princeton, Princeton University Press.

Solé, R. V., Corominas-Murtra, B., and Fortuny, J. 2010. Diversity, competition, extinction: The ecophysics of language change. *Journal of the Royal Society Interface* 7, 1647–64.

Solé, R. V., Corominas-Murtra, B., Valverde, S., and Steels, L. 2010a. Language networks: Their structure, function and evolution. *Complexity* 15, 20–26.

Solé, R. V., and Deisboeck T. S. 2004. An error catastrophe in cancer? *Journal of Theoretical Biology* 228, 47–54.

Solé, R. V., and Goodwin, B. C. 2001. *Signs of Life: How Complexity Pervades Biology.* New York, Basic Books.

Solé, R. V., Manrubia, S. C., Kauffman, S. A., Benton, M., and Bak, P. 1999. Criticality and scaling in evolutionary ecology. *Trends in Ecology and Evolution* 14, 156–60.

Solé, R. V., Manrubia, S. C., Luque, B., Delgado, J., and Bascompte, J. 1996. Phase transitions and complex systems. *Complexity* 1, 13–26.

Solé, R. V., Miramontes, O., and Goodwin, B. C. 1993. Oscillations and chaos in ant societies. *Journal of Theoretical Biology* 161, 343–57.

Solé, R. V., Rodriguez-Caso, C., Deisboeck, T. S., and Saldanya, J. 2008. Cancer stem cells as the engine of unstable tumor progression. *Journal of Theoretical Biology* 253, 629–37.

Solé, R. V., and Valverde, S. 2001. Information transfer and phase transitions in a model of Internet traffic. *Physica* A 289, 595–605.

Sornette, D. 2003. Why Stock Markets Crash. Princeton, Princeton University Press.

Spencer, S. L., Gerety, R. A., Pienta, K. J., and Forrest, S. 2006. Modeling somatic evolution in tumorigenesis. *PLOS Computational Biology* 2, e108.

Stanley, H. E. 1971. *Introduction to Phase Transitions and Critical Phenomena.* New York, Oxford University Press.

Stanley H. E., Amaral, L. A .N., Buldyrev, S. V., Goldberger, A. L., Havlin, S., Leschhorn, H., Maass, P., Makse, H. A., Peng, C. K., Salinger, M. A., Stanley, M. H. R., and Viswanathan, G. M. 2000. Scaling and universality in animate and inanimate systems. *Physica* A 231, 20–48.

Stanley H. E., Buldyrev, S. V., Canpolat, M., Havlin, S., Mishima, O., Sadr-Lahijany, M. R., Scala, A., and Starr, F. W. 1999. The puzzle of liquid water: A very complex fluid. *Physica* D 133, 453–62.

Stauffer, D. 2008. Social applications of two-dimensional Ising models. *American Journal of Physics* 76, 470–73.

Stauffer, D., and Aharony, A. 1992. *Introduction to Percolation Theory*, 2nd ed. London, Taylor and Francis.

Stauffer, D., Schulze, C., Lima, F. W. S., Wichmann, S., and Solomon, S. 2006. Nonequilibrium and irreversible simulation of competition among languages. *Physica* A 371, 719–24.

Stephens, P. A., Sutherland, W. J., and Freckleton, R. P. 1999. What is the Allee effect? *Oikos* 87, 185–90.

Steyn-Ross, D. A., and Steyn-Ross, M. 2010. *Modeling Phase Transitions in the Brain*. New York, Springer.

Strocchi, F. 2005. *Symmetry Breaking*. Berlin, Springer.

Strogatz, S. 1994. *Nonlinear Dynamics and Chaos*. New York, Addison Wesley.

——— 2003. *Sync: How Order Emerges from Chaos in the Universe, Nature, and Daily Life*. New York, Hyperion.

Sumpter, D. 2006. The principles of collective animal behaviour. *Philosophical Transaction of the Royal Society London* 361, 5–22.

Szathmáry, E. 2006. The origin of replicators and reproducers. *Philosophical Transaction of the Royal Society London* 361, 1761–76.

Tainter, J. A. 1988. *The Collapse of Complex Societies*. Cambridge UK, Cambridge University Press.

Takayasu, M., Takayasu, H., and Fukuda, K. 2000. Dynamic phase transition observed in the internet traffic flow. *Physica* A 277, 248–55.

Terborgh, J., et al. 2001. Ecological meltdown in predator-free forest fragments. *Science* 294, 1923–26.

Theraulaz, G., Gautrais, J., Camazine, S., and Deneubourg, J. L. 2003. The formation of spatial patterns in social insects: From simple behaviours to complex structures. *Philosophical Transactions of the Royal Society* A, 361, 1263–82.

Thom, R. 1972. *Structural stability and morphogenesis*: An outline of a general theory of models. Reading, MA, Addison-Wesley.

Tlusty, T. A. 2007. A model for the emergence of the genetic code as a transition in a noisy information channel. *Journal of Theoretical Biology* 249, 331–42.

Tome, T., and Drugowich, J. R. 1996. Probabilistic cellular automaton describing a biological immune system. *Physical Review E* 53, 3976–81.

Toner, J., and Tu, Y. 1998. Flocks, herds, and schools: A quantitative theory of flocking. *Physical Review E* 58, 4828–58.

Turchin, P. 2003. *Historical Dynamics: Why States Rise and Fall.* Princeton, Princeton University Press.

——— 2006. *War and Peace and War: The Life Cycles of Imperial Nations.* New York, Pi Press.

Turner, J. S. 1977. Discrete simulation methods for chemical kinetics. *Journal of Physical Chemistry* 81, 2379–408.

Ujhelyi, M. 1996. Is there any intermediate stage between animal communication and language? *Journal of Theoretical Biology* 180, 71–76.

Valverde, S., Corominas-Murtra, B., Perna, A., Kuntz, P., Theraulaz, G., and Solé, R. V. 2009. Percolation in insect nest networks: Evidence for optimal wiring. *Physics Review E* 79, 066106.

Valverde, S., and Solé, R. V. 2002. Self-organized critical traffic in parallel computer networks. *Physica A* 312, 636–48.

——— 2004. Internet's critical path horizon. *European Physics Journal* B 38, 245–52.

Van Nes E. H., and Schefler, M. 2007. Slow recovery from perturbations as a generic indicator of a nearby catastrophic shift. *American Naturalist.* 169, 738–47.

Vicsek, T., ed. 2001. *Fluctuations and Scaling in Biology.* Oxford, Oxford University Press.

Vicsek, T., Czirók, A., Ben-Jacob, E., Cohen, I., and Shochet, O. 1995. Novel type of phase transition in a system of self-driven particles. *Physical Review Letters* 75, 1226–29.

Wakano, Y. J., Maenosono, S., Komoto, A., Eiha, N., and Yamaguchi, Y. 2003. Self-organized pattern formation of a bacteria colony modeled by a reaction diffusion system and nucleation theory. *Physical Review Letters* 90, 258102.

Wang, W. S.-Y., Ke, J., and Minett, J. W. 2004. Computational studies of language evolution. *Language and Linguistics Monograph Series B* 65108.

Wang, W., and Minett, J. W. 2005. The invasion of language: Emergence, change and death. *Trends in Ecology and Evaluation* 20, 263269.

Watts, D. J., and Strogatz, S. H. 1998. Collective dynamics of "small-world" networks. *Nature* 393, 440–42.

Weidlich, W. 2000. *Sociodynamics: A Systematic Approach to Mathematical Modelling in the Social Sciences*. Amsterdam, Hardwoon Publishers.

Weinberg, R. A. 2007. *Biology of Cancer*. New York, Garland Science.

Wilcove, D. S. 1987. From fragmentation to extinction. *Natural Areas Journal* 7, 23–29.

Wilke, C. O. 2001a. Selection for fitness vs. selection for robustness in RNA secondary structure folding. *Evolution* 55, 2412–20.

——— 2001b. Adaptive evolution on neutral networks. *Bulletin of Mathematical Biology*. 63, 715–30.

Wilke, C. O., Wang, J., Ofria, C., Lenski, R. E., and Adami, C. 2001. Evolution of digital organisms at high mutation rate leads to survival of the flattest. *Nature* 412, 331–33.

Wilson, E. O. 1971. *The Insect Societies*. Cambridge, MA, Harvard University Press.

Wodarz, D., and Komarova, N. 2005. *Computational Biology of Cancer*. Singapore, World Scientific.

Wray, M. A., ed. 2002. *The Transition to Language*. Oxford, UK, Oxford University Press.

Wuensche, A., and Lesser, M. 2000. *The Global Dynamics of Celullar Automata*. Santa Fe, NM, Discrete Dynamics Incorporated.

Yeomans, J. M. 1992. *Statistical Mechanics of Phase Transitions*. Oxford, Oxford University Press.

Yoffee, N., and Cowgill, G. L., eds. 1988. *The Collapse of Ancient States and Civilizations*. Tucson, AZ, University of Arizona Press.

Yook, S.-H, Jeong, H., and Barabási, A.-L. 2002. Modeling the Internet's large-scale topology. *Proceedings of the National Academy of Sciences USA* 99, 13382–86.

Zanette, D. H., and Kuperman, M. N. 2002. Effects of immunization on small-world epidemics. *Physica* A 309, 445–52.

Zanette, D. H. 2008. Analytical approach to bit-string models of language evolution. *International Journal of Modern Physics* C 19, 569–81.

Zeeman, E. C. 1977. Catastrophe theory. *Scientific American* 4, 65–83.

INDEX

A page number in italics refers to a figure or caption

comets, prebiotic molecules derived
from, 70
communication, in random
networks, 67–68. *See also* networks
competition: of chemical species, 72;
dynamics of societies and, 187; of
languages, 173–77, *177*; in tumor
growth, 120, 133
complex systems, 2; cancer as, 121;
ecosystems as, 134; power laws
associated with, 16–17; stable
organization of, 25
computation, phase transitions in, 10
computer networks: mean field model
of, 150–54, *153, 154*, 155;
observed fluctuations in, 148–49,
149, 154; routing tables and,
155–56, 156n; self-organized
critical state in, 154–55; simplified
model of, 150, *151*
computer viruses, 108
connectivity: in computer networks,
150, 155; critical, of random
homogeneous networks, 63,
65–68, *68*; in random Boolean
networks, *114*, 115–17; of
scale-free networks, 69. *See also*
percolation
conservation laws, dimensional
reduction with, *29*, 30
contact process, 100–102, *101*
control parameter, 11; catastrophes
and, 44, *47*; in nonequilibrium
system, 24; for percolation in
lattice, 56; second-order phase
transitions and, 44; temperature as,
3, 11, 15
cooperation: dynamics of societies
and, 187; in ecology of tumors,

133; in multicellular systems, 121;
in prebiotic reaction networks,
71–75, *73*, 76; quadratic
component associated with, 175.
See also swarm behavior
coordination number, of Bethe
lattice, 60, 61
correlation length, 17–18
cosmological evolution, 10
coupling constant, of Ising model,
12, 13
critical boundary: for extinction
under habitat loss, 144, *144*; of
fluid traffic vs. congestion, 153,
154, *154*; in food search by ants,
160, 161; of green-desert
transition, 139, *139, 141*;
of microtubule behavior, 97, *97*
critical connectivity, of random
homogeneous networks, 63,
65–68, *68*
critical dimension, of mean field
approach, 23
critical exponents, 17–18
critical mutation rate, of quasispecies
model, 83, *84*, 86–88
critical phase transitions, 11, 18.
See also phase transitions
critical point, 11
critical slowing down, 49–51, *50*; in
ant colony activity model, 164–65;
ecosystem shifts and, 147
critical temperature of Ising model,
15, *15*; clusters at, *16*; critical
exponents and, 17–18; exact
solution of, 19
cut-off, of real network, 69
cytoskeleton, 91, 92n, 97–98.
See also microtubules

self-organized criticality, 154–55
semiarid ecosystems, 45, 134–35,
   137
sexual contact networks, 108
SIS model, 102–5, 104, 105
smallpox, 99, 107
social collapse, 179–86; factors
   predisposing to, 186; historical
   examples of, 179–81, 180, 186;
   resource depletion and, 181–86,
   184, 185
social insect behavior, 157–66;
   broken symmetry in, 158, 158–61,
   160, 162; order-disorder
   transitions in colonies and,
   162–65, 163, 165; overview of
   complexity in, 157–58, 165–66
social interaction networks, 107, 108
social systems: catastrophes in, 44;
   Ising model of, 11; modeling of,
   187
sol-gel transition, 57–59, 58, 98
spatial correlations: in ecosystems,
   147; in epidemic spreading,
   101–2; ignoring, 23, 63, 68
   (see also mean field models)
spatial dynamics: fitness landscapes
   and, 89; of infectious diseases, 107;
   of tumor growth, 120, 133
spatial segregation, in human
   societies, 187
spatial structures: broken symmetries
   and, 52; percolation and, 166; of
   vegetation, 136–37, 136n, 137.
   See also patterns
spins, in Ising model, 12–13
stability, 25, 26, 32–36
stable equilibrium points, 25, 26, 33,
   34; of logistic model, 36; as

minima of potential, 42, 43.
   See also fixed points
stem-cell switches, 119
sunk-cost effect, 181–82; model
   based on, 182–86, 184, 185
superstring models, 177
surfaces, growing, 10
survival of the flattest, 90
sustainable development, 184,
   184–86, 185
swarm behavior: in fish schools, 8,
   9–10; in insects (see social insect
   behavior). See also cooperation
Swetina-Schuster quasispecies model,
   84–88, 85, 88
symmetry, and phases of matter,
   2–3
symmetry breaking, 38–43, 39, 41; in
   ant colonies, 158, 158–61, 160,
   162; applications of, 52; critical
   slowing down and, 49–51, 50;
   fluctuations close to, 43, 44; in
   Ising model, 39, 41; multiple steps
   of, 51–52, 52. See also bifurcations
system: defined, 53; threshold for
   existence of, 56 (see also percolation
   threshold)
system of linear differential equations,
   29–30
systems biology, 111

Taylor expansion, 34, 34n
technological evolution, 38
temperature: as control parameter, 3,
   11, 15; critical, of Ising model, 15,
   15, 16, 17–18, 19; phases of
   matter and, 2–3
tensegrity, 97–98
termites, 158, 166